Essential Oils - Recent Advances, New Perspectives and Applications

Edited by Jonas Viškelis

Published in London, United Kingdom

Essential Oils - Recent Advances, New Perspectives and Applications
http://dx.doi.org/10.5772/intechopen.107782
Edited by Jonas Viškelis

Contributors
Gullelaala Khan, Gulshan Irshad, Farah Naz, Raina Ijaz, Abd-ur-Rehman Khalid, Muhammad Usman Raja, Sajid Mehmood, Nayla Haneef, Amar Mehmood, Nagina Rafiq, Saiqa Bashir, Amna Maqsood, Liliana Betancourt López, Imen Dridi, Ahmed Landoulsi, Nadia Smirani, Kolajo Adedamola Akinyede, Habeebat Adekilekun Oyewusi, Oluwatosin Olubunmi Oladipo, Oladimeji Samuel Tugbobo, Saber Jedidi, Hichem Sebai, Ibrahim Mohamed Khattab, Mohamed Fathy Elgandy, Zoubeida Souiy, Mkaddem Mounira, Elena Stashenko, Jairo René Martínez, Mimouna Yakoubi, Nasser Belboukhari, Khaled Sekkoum, Mohammed Bouchekara, Michael Ibok, Oluwakayode Odeja, Ejike Onwudiegwu Okpala

First published in London, United Kingdom, 2024 by IntechOpen
IntechOpen is the global imprint of INTECHOPEN LIMITED, registered in England and Wales, registration number: 11086078, 5 Princes Gate Court, London, SW7 2QJ, United Kingdom
Printed in Croatia

British Library Cataloguing-in-Publication Data
A catalogue record for this book is available from the British Library

Additional hard and PDF copies can be obtained from orders@intechopen.com

Essential Oils - Recent Advances, New Perspectives and Applications
Edited by Jonas Viškelis
p. cm.

This title is part of the Biochemistry Book Series, Volume 49
Topic: Chemical Biology
Series Editor: Miroslav Blumenberg
Topic Editors: Şükrü Beydemir and Deniz Ekinci

Print ISBN 978-0-85014-204-4
Online ISBN 978-0-85014-205-1
eBook (PDF) ISBN 978-0-85014-206-8
ISSN 2632-0983

For EU product safety concerns:
IN TECH d.o.o., Prolaz Marije Krucifikse Kozulić 3, 51000 Rijeka, Croatia,
info@intechopen.com or visit our website at intechopen.com.

IntechOpen Book Series
Biochemistry
Volume 49

Aims and Scope of the Series

Biochemistry, the study of chemical transformations occurring within living organisms, impacts all of the life sciences, from molecular crystallography and genetics, to ecology, medicine and population biology. Biochemistry studies macromolecules - proteins, nucleic acids, carbohydrates and lipids –their building blocks, structures, functions and interactions. Much of biochemistry is devoted to enzymes, proteins that catalyze chemical reactions, enzyme structures, mechanisms of action and their roles within cells. Biochemistry also studies small signaling molecules, coenzymes, inhibitors, vitamins and hormones, which play roles in the life process. Biochemical experimentation, besides coopting the methods of classical chemistry, e.g., chromatography, adopted new techniques, e.g., X-ray diffraction, electron microscopy, NMR, radioisotopes, and developed sophisticated microbial genetic tools, e.g., auxotroph mutants and their revertants, fermentation, etc. More recently, biochemistry embraced the 'big data' omics systems. Initial biochemical studies have been exclusively analytic: dissecting, purifying and examining individual components of a biological system; in exemplary words of Efraim Racker, (1913 –1991) "Don't waste clean thinking on dirty enzymes." Today, however, biochemistry is becoming more agglomerative and comprehensive, setting out to integrate and describe fully a particular biological system. The 'big data' metabolomics can define the complement of small molecules, e.g., in a soil or biofilm sample; proteomics can distinguish all the proteins comprising e.g., serum; metagenomics can identify all the genes in a complex environment e.g., the bovine rumen.
This Biochemistry Series will address both the current research on biomolecules, and the emerging trends with great promise.

Meet the Series Editor

Miroslav Blumenberg, Ph.D., was born in Subotica and received his BSc in Belgrade, Yugoslavia. He completed his Ph.D. at MIT in Organic Chemistry; he followed up his Ph.D. with two postdoctoral study periods at Stanford University. Since 1983, he has been a faculty member of the RO Perelman Department of Dermatology, NYU School of Medicine, where he is codirector of a training grant in cutaneous biology. Dr. Blumenberg's research is focused on the epidermis, expression of keratin genes, transcription profiling, keratinocyte differentiation, inflammatory diseases and cancers, and most recently the effects of the microbiome on the skin. He has published more than 100 peer-reviewed research articles and graduated numerous Ph.D. and postdoctoral students.

Meet the Volume Editor

Dr. Jonas Viškelis is a chief researcher and head of the Biochemistry and Technology laboratory at the Institute of Horticulture, Lithuanian Research Centre for Agriculture and Forestry (LAMMC). He defended his dissertation in 2018. He has seventy-six scientific publications to his credit, sixty-two of which are in the Clarivate Analytic WOS database with a citation index. He has presented his research at more than 100 scientific conferences and implemented more than 40 R&D projects, including 4 international projects (2 for Eureka and 2 for Interreg Europe). In 2016, 2017, and 2018 he received a doctoral scholarship for academic achievements awarded by the Lithuanian Council of Science. In 2019, Dr. Viškelis won the Research Council of Lithuania competition for young scientists. In 2021 he was elected a member of the Young Academy of the Lithuanian Academy of Sciences.

Contents

Preface

In recent years, essential oils have gained significant attention and recognition for their diverse range of benefits and applications.

The use of essential oils has continued to grow in popularity across various industries, including medicine, agriculture, food, and cosmetics. Their therapeutic properties and natural origins make them a valuable resource for a wide range of applications. These oils, extracted from aromatic plants through steam or hydro-distillation, have a long history of use in ancient civilizations for various purposes such as pain management, wound care, respiratory tract complaints, aromatherapy, and spiritual relaxation. Furthermore, essential oils serve as a rich source of compounds that are used in different fields. Essential oils have become increasingly important in scientific research and industrial settings, from nutritional and pharmaceutical uses to perfumery and herbal beverages.

The diverse range of therapeutic effects of essential oils is a testament to the varied bio-benefits that aromatic plants offer. With approximately 3000 essential oils known, and 300 of them being commercially important, the potential for their application in the pharmaceutical, agronomic, food, cosmetic, and perfume industries is vast and continually expanding. The extraction techniques used to obtain essential oils are vital in preserving the bioactive compounds that contribute to their therapeutic properties. Each plant requires specific extraction methods to ensure the highest quality of essential oils. These oils have found applications in a wide array of products, including food, drinks, perfumes, pharmaceuticals, and cosmetics, due to their rich and complex chemical composition.

Furthermore, essential oils' historical significance and continued relevance in modern times demonstrate their enduring appeal. The ancient civilizations' use of essential oils for pain management, wound care, and spiritual relaxation underscores their time-tested efficacy and versatility. As the scientific community delves deeper into the potential of essential oils, their significance as a valuable natural resource will only continue to grow.

Jonas Viškelis
Institute of Horticulture,
Lithuanian Research Centre for Agriculture and Forestry,
Babtai, Lithuania

Section 1

Essential Oils Biochemistry

Chapter 1

Phytochemistry, Medicinal Uses, and Beneficial Nutritional Effects of Essential Oils

Saber Jedidi and Hichem Sebai

Abstract

Plants contain a considerable reservoir of secondary metabolites (flavonoids, tannins, and essential oils). These molecules exhibit variations in chemical structure as well as a very wide range of biological activities. Essential oils (EOs) are secondary metabolites produced by aromatic plants. EOs contain bioactive molecules, mainly represented by monoterpene hydrocarbons, oxygenated monoterpenes and sesquiterpenes, and sesquiterpene hydrocarbons. The organoleptic properties and biological activities of EOs are distinguished by their respective compositions. They have long been recognized for their medicinal properties such as antibacterial, antifungal, bio-herbicide, antioxidant, anti-inflammatory, antidiabetic, and hepato-nephroprotective activities. These organic compounds also exert beneficial effects on the nutrition of ruminants, by modulating digestibility and reducing the emission of methane, a greenhouse gas. This chapter is devoted to the study of chemical composition, medicinal uses, and beneficial nutritional effects of essential oils.

Keywords: essential oils, extraction method, phytochemistry, therapeutic potential, nutritional effects

1. Introduction

In recent decades, an increased interest has been given to alternative medicine. In this respect, the return to herbal care is highly recommended. Indeed, plants and their therapeutically active substances have been widely used in ethno-medicine [1]. On the other hand, researches on the benefits of physiotherapy and aromatherapy, using essential oils (EOs) for healing purposes, were constantly increasing [2]. The medicinal properties of essential oils are thus widely described.

Essential oils (EOs) are products with a rather complex composition, containing volatile active ingredients. Physically, these are volatile, which differentiates them from fixed oils. They are colorless liquids with a generally strong odor and flavor [3]. They are poorly miscible with water and well soluble in oils and organic solvents [4]. Finally, the obtained oils are separated by the difference in density, generally by simple decantation [5].

According to botanists, there are approximately 800,000 to 1500,000 plant species, 10% of which contain EOs. Indeed, they have been reported to be present in about 2000 species distributed in 60 botanical families, such as Lamiaceae, Lauraceae Myrtaceae, Rutaceae, Asteraceae, Cupressaceae, Poaceae, and Zingiberaceae [6].

EOs are located in various parts of the plant (roots, fruits, seeds, flowers, leaves, bark, and wood). The biosynthesis of these compounds realizes in plant cells via various metabolic reactions, such as isopentenyl diphosphate and its isomer dimethylallyl diphosphate. The end products of EOs are terpenoid and are synthesized with a large group of enzymes called terpene synthases. They occur in the cytoplasms of plants. Then, they localize in several organs of the plant such as trichomes, epidermal cells, and, finally, the secretory pockets [7].

According to reprts, EOs are distinguished by their smells, colors, densities, and chemotypes. Therefore, each essential oil has its characteristics, fragrance, and properties. Importantly, EOs are defined by their botanical species, part of the plant, the extraction mode, and the characteristic active principle [8].

EOs are characterized with several biological activities, such as fungicide, insecticide, herbicide, and bactericide potentials. These could be used as antiseptic and antimicrobial properties [9]. Many reports have shown that these volatile compounds have antioxidant, antiviral, and antiparasitic properties. EOs are also utilized as drugs in cancer chemotherapy [10]. These bioactive molecules have been useful in dentistry for the disinfection of dental pulp and the treatment/prevention of caries [11].

In animal nutrition, essential oils have attracted the attention of nutritionists for their potential role as an alternative to growth-promoting antibiotics. In small ruminants, essential oils are characterized for their beneficial effects on the digestion and digestibility of food. More importantly, research has shown that the inclusion of essential oils at reasonable doses modulated rumen fermentation parameters such as organic matter digestibility (OMD), volatile fatty acids (VFA), and metabolizable energy (ME) [12, 13].

On the other hand, human and animal poisoning with essential oils has been reported in several research works [14, 15]. Due to the increase uses of essential oils, the number of poisonings is expected to amplify in the future. It is therefore interesting to use EOs in animals and humans with caution.

In the available chapter, we firstly defined the essential oils, by studying their extraction methods and chemical structure, as well as their main uses. Then, in a second part, we developed their pharmacological and beneficial effects in animal nutrition.

2. Methods of essential oils extraction

Several methods of EOs extraction have been previously described. This diversity is due to the plant material variety and the sensitivity of their constituents. The choice of mode depends on the nature and the plant material parts, the physicochemical characteristics of EOs, as well as the extract uses. In this chapter, we will present some techniques for the EOs extraction.

2.1 Distillation by steam entrainment

At the Sylvo-Pastoral Resources laboratory in Tabraka, the technique consists of distilling 10 to 15 kg of plant material in the alembic with 10 L of water separated by a grid in a still (**Figure 1**). The distillation temperature is 100°C. The operation leads to the release of water vapor, which plays a dual role: (i) release of the plant essence

Figure 1.
Extraction of essential oils from Pinus halepensis needles by steam distillation. A: Water filling up to separation grid level; B: Introduction of plant material in the alembic. C: Complete assembly.

and (ii) transport of the oil, which condenses during cooling. The obtained liquid contains a fraction of essential oil and another of floral water. These two constituents are separated by the difference in density.

2.2 Hydrodistillation

This technique is the oldest method used. In brief, the plant material is immersed directly in a still filled with water placed on a heat source. The mixture is then brought to a boil. The vapors are condensed in a cooler (**Figure 2**). These two obtained constituents are separated by the difference in density [17].

2.3 Hydrodiffusion

It involves spraying water vapor through plant material in a top-down position. In this case, the vapor passing through the plant material is downward, as opposed to other traditional distillation techniques (**Figure 3**). This technique is more advantageous, since it qualitatively and quantitatively improves the harvested essential oil, saving time, steam, and energy [18].

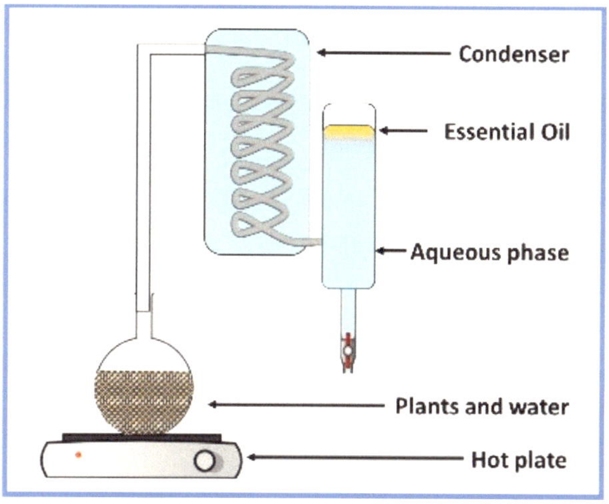

Figure 2.
Extraction of essential oils by hydrodistillation [16].

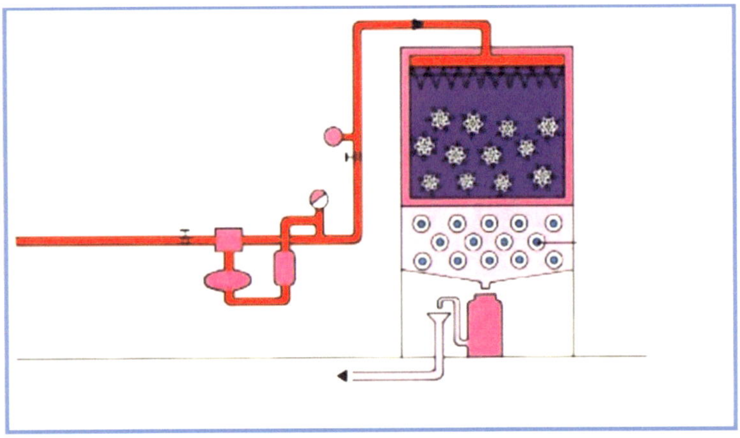

Figure 3.
Extraction of essential oils by hydrodiffusion.

2.4 Expression

This technique of expression or cold pressing essentially concerns the extraction of essential oils from the zest of lemons and oranges. Finally, the essence is released by a stream of water, then decanted (**Figure 4**). A new mechanical technique based on the bursting of the oleiferous bags under the effect of either a depression or abrasion of the fresh bark would eliminate water and reduce the effects of the essences compounds oxidation [20].

2.5 Solvent extraction

The solvent extraction technique consists of placing a volatile solvent and the plant material in an extractor. The product thus obtained is called "concrete". This concrete can then be brewed with absolute alcohol, filtered, and iced to extract the vegetable waxes.

After a final concentration, an "absolute" is obtained.

Among the features of this technique are:

- Yields are generally higher compared to conventional distillation.
- The use of organic solvents, which can lead to risks of artifacts and possibilities of contamination of the sample by impurities that are sometimes difficult to eliminate [21].

2.6 Microwave extraction

This technique is called Solvent Free Microwaves Extraction. It consists of extracting the essential oil using constant energy microwave radiation and a vacuum sequence. This technology is a combination of microwave heating and atmospheric pressure distillation. This method consists of introducing the part of the plant to be extracted into microwave reactor, without organic solvent/water. The increase in the

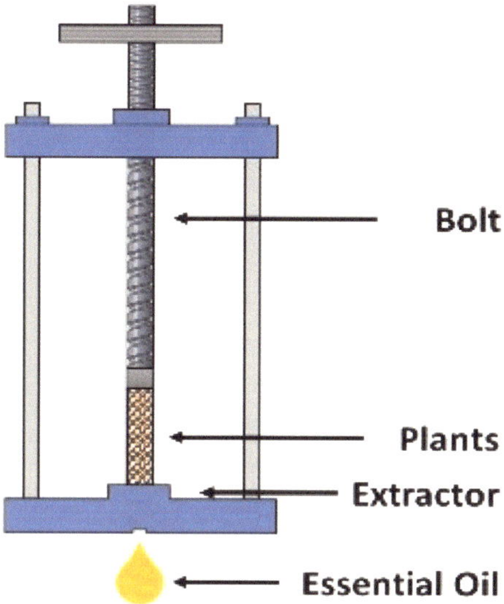

Figure 4.
Cold pressing method [19].

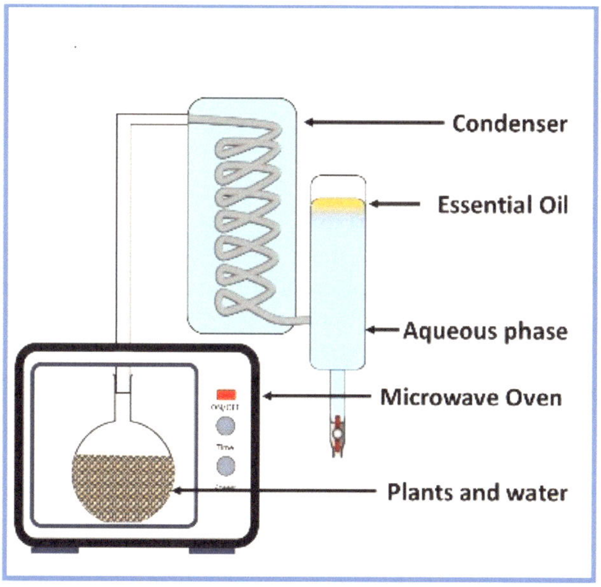

Figure 5.
Microwave-assisted hydrodistillation [16].

vegetable aqueous fraction allows the rupture of the glands containing the essential oil. Finally, a cooling system outside the microwave allows the condensation of the distillate (**Figure 5**). This process seems to be more competitive and economical than the conventional methods [22].

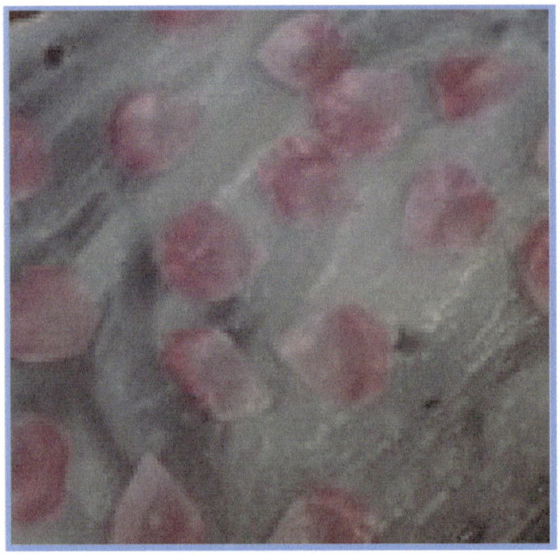

Figure 6.
Enfleurage of rose petals.

2.7 Extraction by fatty substances

This enfleurage technique is used in the extraction of EO from flowers, fragile parts of plants. It takes the advantage of the fat-soluble odorous components of plants in fatty substances. The principle consists of bringing the flowers into contact with a fatty substance to saturate it with plant essence (**Figure 6**). The obtained product is a floral ointment that is then exhausted by a solvent, which is eliminated under reduced pressure [23].

3. Chemical structure of essential oils

The chemical composition of EO can be identified by chromatographic analysis such as gas chromatography and mass spectrometry (GC/MS) [12]. EO are complex and variable mixtures of constituents that generally belong to two groups characterized by different biogenetic origins, the groups of terpenoids and aromatic compound derived from phenylpropane, which are much less frequent.

3.1 Terpenoids

Terpenes are formed from n multiples of the C_5H_8 isoprene; they are multicyclic structures that differ not only in the functional groups but also in the basic structure of their hydrocarbon skeletons. When n = 2, the terpene corresponds to monoterpenes $(C_{10}H_{16})_2$. Monoterpenes and their derivatives are therefore linear chains or cycles formed from two isoprene units.

3.1.1 Monoterpenes

Carbides are almost always present. They are acyclic, monocyclic, or bicyclic. They sometimes constitute more than 90% of the essential oil (**Figure 7**).

Figure 7.
Acyclic (myrcene) and monocyclic (thymol) monoterpene [3].

3.1.2 Sesquiterpenes

They consist of 3 isoprene units. Chain elongation increases the number of possible cyclizations (**Figure 8**).

3.1.3 Other terpenoids

These subgroups contain more than three isoprene units. They are presented as follows:

- 4 isoprene units: diterpenoids,
- 5 isoprene units: sesterterpenoids,
- 6 isoprene units: triterpenoids,
- 8 isoprene units: tetraterpenoids,
- Compounds whose number exceeds 8 isoprene units: polyterpenoids

3.2 Aromatic compounds

Phenylpropane derivatives (C_6–C_3) are much less common than the previous ones. Very frequently, these are allyl and propenylphenols, sometimes aldehydes, characteristics of certain essential oils, such as that of clove (eugenol). On the other hand, the safarole is a compound with the chemical formula C_6-C_1 has been rarely identified in essential oils composition [24].

3.3 Compounds of various origins

Depending on their mode of extraction, essential oils can contain various aliphatic compounds, generally of low molecular mass, which can be carried away during

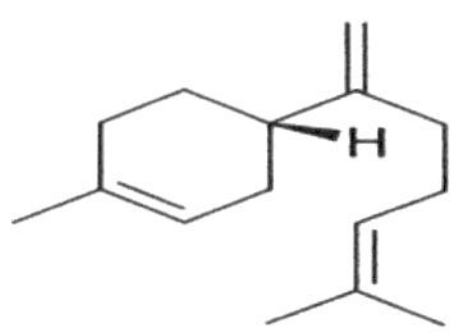

Figure 8.
Sesquiterpene (β-besabolene) [3].

hydrodistillation, such as carbide, acid (C3 to C10), alcohols, aldehydes (octanal, decanal), esters, lactones, nitrogen, or sulfur products [25].

4. Biological activities of EO

The biological activity of EOs depends on their chemical compositions and structures and on the synergistic effects between their major and minor compounds.

4.1 Antioxidant capacity

The antioxidant activity was evaluated, *in vitro*, by various tests. In this respect, Louail et al. [26] assessed antioxidant activities of *Ammodaucus leucotrichus* EO by measuring by inhibiting the β-carotene bleaching. The value of the essential oils showed a better antioxidant activity when compared to ascorbic acid, used as reference antioxidant molecule. In another experiments, Selmi et al. [13], Aloui et al. [27], and Jedidi et al. [28] assessed the antioxidant activities of essential oils obtained from aerial parts of *Junepeus phoenicea*, *Pinus halepensis*, and *Rosmarinus officinalis*, using 2, 2-diphenyl-1-picrylhydrazyl (DPPH$^{\bullet}$) and 2,2′-azino-bis [3-ethylbenzthiazoline-6-sulphonic acid] (ABTS) tests. Essential oils of the four species exhibited strong antioxidant abilities to reduce the studied radicals. It has been demonstared that 1,8-cineole, α-pinene and camphor have the dominant components of the EO of a few commercial species. These identified molecules are powerful scavengers of free radicals and therefore responsible for the strong antioxidant activities observed [29].

4.2 Antimicrobial activity

The EO of many species were screened for their antimicrobial activities against different microorganisms, including Gram-positive and Gram-negative bacteria. Essentail oils exhibited strong inhibitory action against most tested organisms. Furthermore, the essential oil showed significant antibacterial activity against Gram-negative and Gram-positive bacteria. It has been shown that Gram (+) bacteria are more sensitive to essential oils isolated from the species *Pinus pinea* [27], *Ammodaucus leucotrichus* [30], and *Rosmarinus officinalis* [31]. The authors suggested that this finding can be explained by the fact that Gram (−) bacteria have hydrophobic lipopolysaccharide in the outer part of their membranes, which provides effective protection against different agents.

In addition, listerine, which is a solution consisting of thymol and eucalyptol essential oils, has a high bactericidal activity on microorganisms in saliva and dental plaque [32].

On the other hand, it also demonstrated that the sesquiterpenoids identified in the essential oils of *Cyperus iria* leaves exerted a potential fungicidal action against Fusarium graminearum [33].

4.3 Anticholinesterase activity

Limonene, a monoterpene isolated from Ammodaucus leucotrichus essential oils, has been shown to induce high acetylcholinesterase inhibitory activity with an IC_{50} of about 51.6 $\mu g\ ml^{-1}$ [34]. Additionally, Aazza et al. [29] proved that 1,8-cineole, α-pinene, and camphor were the dominant components of sage essential oils and were responsible for strong antiacetylcholinesterase activity.

5. Medicinal uses

5.1 Antiulcer ptentiel

Essential oils are widely used in the treatment/prevention of digestive pathologies. Terpenes and phenylpropanoids found in many essential oils have been shown to have potential for use in peptic ulcer disease [35]. The authors suggested that the anti-ulcer action is due to bioactive volatile molecules and their mechanisms of action, such as restoring the activity of antioxidant enzymes and the level of nonenzymatic antioxidants, recovering the level of mediators intracellular, and restorating pH. Essential oils also exerted an effect against *Helicobacter pylori* bacteria and improved the gastric mucosal barrier.

5.2 Antidiabetic capacity

EOs have shown a series of biological properties with health-promoting conduct in humans. Among the chemical groups that make up essential oils, terpenoids are characterized by their hypoglycemic effect. These molecules inhibit enzymes responsible for the development of insulin resistance and normalization of plasma glucose and insulin levels [36]. These bioactive molecules exert an inhibitory effect in the process of carbohydrate metabolism and prevent the phenomenon of insulin resistance and finally of the serum glycaemia level [37]. In addition, it has been suggested that triterpenes have contributed to the treatment of diabetic neuropathy and nephropathy by inhibiting several pathways involved in the diabetes and associated complications [36]. On the other hand, the *Aegle marmelos* leaves' volatile molecules showed strong antidiabetic activity [38]. In another report, it was demonstared that the antidiabetic effetct of *Lavandula stoechas* essential oil is due, partly, to its potent antioxidant properties [39].

5.3 Anti-inflammatory properties

The different compounds of essential oils are also characterized by powerful anti-inflammatory effects. In this context, Baricevic et al. [40] proved that ursolic acid isolated from the *Salvia officinalis* EO was the main component of its anti-inflammatory power. On the other hand, certain terpenoids such as scropolioside like iridoids have shown potential for anti-inflammatory, hepatoprotective, and wound-healing activity [41, 42]. Terpenes and terpenoids are promising in the treatment intestinal inflammatory and have been also shown to display a broad range of biological activities in various human disease models [43]. Few other volatile terpenes and terpenoids, mainly monoterpenes, oxygenated terpenes, terpene esters, and sesquiterpenes, showed anti-inflammatory properties [44]. In addition, the anti-inflammatory effect of the *Ammodaucus leucotrichus* fruits' EOs has been demonstrated by the evaluation of the antiedematogenic response of essential oils in Carrageenan-induced hind paw edema in animal model [45].

5.4 Anticancer activity

In vitro, certain terpenoids isolated from the roots of *Salvia officinalis* exerted a protective effect against cellular and DNA damage during human carcinomas. On the other hand, the α-humulene, a sesquiterpene compound identified in officinal sage essentail oils, demonstrated strong cytotoxic activity in human prostate carcinoma cells. Trans-caryophyllene, isolated from *Salvia officinalis*, also exerted high cytotoxic activity against renal cell carcinoma cells [46, 47]. In addition, research work conducted by Li et al. [48] showed that natural bicyclic sesquiterpenes exhibited potential anticancer activity.

5.5 Hepato-nephroprotective effect

Volatile compounds are known for their hepato-hepato-nephroprotective actions. In this regard, Fahmy et al. [49] evaluated, on animal model, the potential effect of essential oils extracted from sage plant against carbon tetrachloride (CCl4)-induced hepato/renal toxicities. This study confirmed that *Salvia officinalis* essential oils (SOEO) represent a potential candidate to reverse CCl_4-associated hepato/renal damage. This effect may occur via an antioxidant defense mechanism that is partly related to the complexity of its chemical constituents.

6. Beneficial nutritional effects of EO

6.1 In broiler chickens

EOs are natural bioactive molecules that can be included as alternatives to antibiotics during broiler rearing. Results published by Puvača et al. [50] confirmed that these compounds have beneficial effects on nutrient digestibility, microbiota, and gut function. Authors also suggested that EOs have positive effects, but knowledge of their use in poultry feed is still limited and requires further research.

6.2 In ruminants

EOs could be used as additives in ruminant feed, to modify the activity of microorganisms responsible for ruminal fermentations and to improve digestion in small ruminants.

Essential oils have been shown to exert beneficial effects on ruminal fermentations. In fact, these molecules are involved in improving the amino acids' quantity available for animal's needs, increasing the volatile fatty acid (VFA) levels, providing sources of energy for animals, and reducing methane and ammonia emissions [51]. These results are similar to those published by Jedidi et al. [12] and Selmi et al. [13] who worked on the effect of essential oils from many plants. These authors found that the inclusion of volatile compounds, at reasonable doses, stimulated parameters of ruminal fermentation in vitro, such as digestibility of organic matter (DOM), metabolisable energy (ME), and volatile fatty acids (VFA). In addition, EOs exerted the decrease in greenhouse gas emissions in a dose-dependent manner. The authors mentioned that all of these beneficial effects of essential oils are attributed to the synergistic effects of these components.

In this respect, a mixture of volatile compounds extracted from several aromaticplants has been shown to significantly improve food bioavailability. This beneficial effect exerted by this preparation is due to the changes in the intestinal ecosystem [52].

In the field of animal production, EOs are mainly used to improve zootechnical performance, such as growth rate, consumption index (CI), feed intake level, feed digestibility, and animal health status [53].

7. Conclusions

This chapter has shown that monoterpenes, sesquiterpenes and their derivatives, aromatic compound, and compound of various origins are the main chemical constituents of essential oils identified in plants. Several reports demonstrated that EOs exhibit a range of pharmacological actions, such as antiulcer, hepato-nephroprotective, anti-inflammatory, antidiabetic, antioxidant, antibacterial, antifungal, and anticholinesterase activities, supporting their traditional uses. In addition, essential oils exerted beneficial effects on nutrient digestibility and intestinal function in broilers, improved digestibility parameters in ruminants, as well as decreased NH_4^+ emissions.

However, further improvements are needed to adjust the doses of inclusion/administration because misuse can lead to severe human and animal toxicities. Finally, it is strongly recommended to keep these bioactive molecules well and to keep them away from children.

Conflict of interest

The authors declare no conflict of interest.

Author details

Saber Jedidi[1,2*] and Hichem Sebai[1]

1 Unit of Functional Physiology and Valorization of Bio-Resources, University of Jendouba, Higher Institute of Biotechnology of Béja, Tunisia

2 Laboratory of Sylvo-Pastoral Resources, Sylvo-Pastoral Institute of Tabarka, University of Jendouba, Tabarka, Tunisia

*Address all correspondence to: jedidi_saber@yahoo.fr

References

[1] Jedidi S, Aloui F, Selmi H, Rtibi K, Dallali S, Sebai H. Ethnobotanical survey on the traditional use of officinal sage (*Salvia officinalis* L.) in Tabarka and Aïn Draham (Northwestern of Tunisia). Journal of New Sciences. 2018;**18**:3402-3412

[2] Dalcin MS, Caf AC, de Almeida SR, do Nascimento IR, de Souza Ferreira TP, de Sousa Aguiar RW, et al. Evaluation of essential oils for preventive or curative management of melon gummy stem blight and plant toxicity. Journal of Medicinal Plants Research. 2017;**11**(26):426-432. DOI: 10.5897/JMPR2017.6405

[3] Sarkic A, Stappen I. Essential oils and their single compounds in cosmetics-a critical review. Cosmetics. 2018;**5**(1):11. DOI: 10.3390/molecules26030666

[4] Barradas TN, de Holanda e Silva KG. Nanoemulsions of essential oils to improve solubility, stability and permeability: A review. Environmental Chemistry Letters. 2021;**19**(2):153-1171. DOI: 10.1007/s10311-020-01142-2

[5] El Asbahani A, Miladi K, Badri W, Sala M, Addi EA, Casabianca H, et al. Essential oils: From extraction to encapsulation. International Journal of Pharmaceutics. 2015;**483**(1-2):220-243. DOI: 10.1016/j.ijpharm.2014.12.069

[6] Moshrefi Zenoozi Z, Soltaninezhad B, Hashemi M, Noori SMA. A review of effective essential oils and their biologically active compounds to protect the safety of food stored against insect pests. Journal of Essential Oil Research. 2022;**34**(2):111-122. DOI: 10.1080/10412905.2022.2032420

[7] Rehman R, Hanif MA, Mushtaq Z, Al-Sadi AM. Biosynthesis of essential oils in aromatic plants: A review. Food Reviews International. 2016;**32**(2):117-160. DOI: 10.1080/87559129.2015.1057841

[8] Besombes C. Contribution à l'étude des phénomènes d'extraction hydro-thermo-mécanique d'herbes aromatiques. Applications généralisées [thesis]. France: Université de La Rochelle; 2008

[9] Mancianti F, Ebani VV. Biological activity of essential oils. Molecules. 2020;**25**(3):678. DOI: 10.3390/molecules25030678

[10] Bhalla Y, Gupta VK, Jaitak V. Anticancer activity of essential oils: A review. Journal of the Science of Food and Agriculture. 2013;**93**(15):3643-3653. DOI: 10.1002/jsfa.6267

[11] Singh I, Kaur P, Kaushal U, Kaur V, Shekhar N. Essential oils in treatment and management of dental diseases. Biointerface Research in Applied Chemistry. 2022;**12**:7267-7286. DOI: 10.33263/BRIAC126.72677286

[12] Jedidi S, Semi H, Aloui F, Dhifallah A, Sammari H, Abbes C, et al. Antioxidant and antibacterial properties of *Salvia officinalis* essential oils and their effects on *In vitro* feed fermentation with goat rumen liquor. Animal Nutrition and Feed Technology. 2022;**22**(2):297-310. DOI: 10.5958/0974-181X.2022.00024.5

[13] Selmi H, Dhifallah A, Jedidi S, Bahri A, Abbes C, Rouissi H. Biological activities of *Juniperus phoenicea* essential oil and impact on in vitro ruminal fermentation in sheep. Large Animal Review. 2023;**29**(2):83-88

[14] Hayes AJ, Markovic B. Toxicity of Australian essential oil Backhousia citriodora (lemon myrtle). Part 1. Antimicrobial activity and in vitro cytotoxicity. Food and Chemical Toxicology. 2002;**40**(4):535-543. DOI: 10.1016/S0278-6915(01)00103-X

[15] Kouninki H, Ngamo LST, Hance T, Ngassoum MB. Potential use of essential oils from local Cameroonian plants for the control of red flour weevil Tribolium castaneum (Herbst.) (Coleoptera: Tenebrionidae). African Journal of Food, Agriculture, Nutrition and Development. 2007;**7**(5):1-15. DOI: 10.18697/ajfand.16.2145

[16] Haro-González JN, Castillo-Herrera GA, Martínez-Velázquez M, Espinosa-Andrews H. Clove essential oil (*Syzygium aromaticum* L. Myrtaceae): Extraction, chemical composition, food applications, and essential bioactivity for human health. Molecules. 2021;**26**(21):6387. DOI: 10.3390/molecules26216387

[17] Sakkas H, Papadopoulou C. Antimicrobial activity of basil, oregano, and thyme essential oils. Journal of Microbiology and Biotechnology. 2017;**27**(3):429-438. DOI: 10.4014/jmb.1608.08024

[18] Bassereau M, Chaintreau A, Duperrex S, Joulain D, Leijs H, Loesing G, et al. GC-MS quantification of suspected volatile allergens in fragrances. 2. Data treatment strategies and method performances. Journal of Agricultural and Food Chemistry. 2007;**55**(1):25-31. DOI: 10.1021/jf062028l

[19] Hatami T, Johner JC, Zabot GL, Meireles MAA. Supercritical fluid extraction assisted by cold pressing from clove buds: Extraction performance, volatile oil composition, and economic evaluation. The Journal of Supercritical Fluids. 2019;**144**:39-47. DOI: 10.1016/j.supflu.2018.10.003

[20] Samuelson R, Lobl M, Higgins S, Clarey D, Wysong A. The effects of lavender essential oil on wound healing: A review of the current evidence. Journal of Alternative and Complementary Medicine. 2020;**26**(8):680-690. DOI: 10.1089/acm.2019.0286

[21] Abbasi H, Rezaei K, Rashidi L. Extraction of essential oils from the seeds of pomegranate using organic solvents and supercritical CO_2. Journal of the American Oil Chemists' Society. 2008;**85**(1):83-89. DOI: 10.1002/aic.10844

[22] Moradi S, Fazlali A, Hamedi H. Microwave-assisted hydro-distillation of essential oil from rosemary: Comparison with traditional distillation. Avicenna Journal of Medical Biotechnology. 2018;**10**(1):22-28

[23] Kanlayavattanakul M, Kitsiripaisarn S, Lourith N. Aroma profiles and preferences of *Jasminum sambac* L. flowers grown in Thailand. Cosmetic Sciences. 2013;**64**(6):483-493

[24] Siva S, Li C, Cui H, Meenatchi V, Lin L. Encapsulation of essential oil components with methyl-beta-cyclodextrin using ultrasonication: Solubility, characterization, DPPH and antibacterial assay. Ultrasonics Sonochemistry. 2020;**64**:104997. DOI: 10.1016/j.ultsonch.2020.104997

[25] Carson CF, Hammer KA. Chemistry and Bioactivity of Essential Oils. Lipids and Essential Oils as Antimicrobial Agents. Reykjavik, Iceland: Wiley Online Library; 2011. pp. 203-238. DOI: 10.1002/9780470976623.ch9

[26] Louail Z, Kameli A, Benabdelkader T, Bouti K, Hamza K, Krimat S. Antimicrobial and antioxidant activity of essential oil of *Ammodaucus leucotrichus* Coss. & Dur. Seeds. Journal of Materials and Environmental Science. 2016;7(7):2328-2334

[27] Aloui F, Baraket M, Jedidi S, Hosni K, Bouchnak R, Salhi O, et al. Chemical composition, anti-radical and antibacterial activities of essential oils from needles of *Pinus halepensis* mill., *P. pinaster* Aiton., and *P. pinea* L. Journal of Essential Oil Bearing Plants. 2021;**24**(3):453-460. DOI: 10.1080/0972060X.2021.1943541

[28] Jedidi S, Aloui F, Selmi H, Jdaidi N, Ghribi S, Sebai H. Phytochemical profile and antioxidant properties of essential oils isolated from the aerial parts of *Rosmarinus officinalis* L. cultivatedin northwestern of Tunisia. Moroccan journal of. Agricultural Sciences. 2022;**3**(4):229-233

[29] Aazza S, Lyoussi B, Miguel MG. Antioxidant and antiacetylcholinesterase activities of some commercial essential oils and their major compounds. Molecules. 2011;**16**(9):7672-7690. DOI: 10.3390/molecules16097672

[30] Dahmane D, Dob T, Krimat S, Nouasri A, Metidji H, Ksouri A. Chemical composition, antioxidant and antibacterial activities of the essential oils of medicinal plant *Ammodaucus leucotrichus* from Algeria. Journal of Essential Oil Research. 2017;**29**(1):48-55. DOI: 10.1080/10412905.2016.1201015

[31] Jedidi S, Aloui F, Selmi H, Rtibi K, Sammari H, Sebai H. Antibacterial capacity and inhibitory effect of rosemary (*Rosmarinus officinalis* L.) essential oils on gastrointestinal transit in rats. Acta Scientific Nutritional Health. 2023;7(1):115-120. DOI: 10.31080/ASNH.2022.07.1176

[32] Goetz P, Ghedira K. Introduction à la phytothérapie anti-infectieuse. In: Phytothérapie anti-infectieuse. Collection Phytothérapie Pratique. Paris: Springer; 2012. pp. 3-14. DOI: 10.1007/978-2-8178-0058-5_1

[33] Jiang Y, Ownley BH, Chen F. Terpenoids from weedy ricefield flatsedge (*Cyperus iria* L.) are developmentally regulated and stress-induced, and have antifungal properties. Molecules. 2018;**23**:3149. DOI: 10.3390/molecules23123149

[34] Sadaoui N, Bec N, Barragan-Montero V, Kadri N, Cuisinier F, Larroque C, et al. The essential oil of Algerian *Ammodaucus leucotrichus* Coss. & Dur. And its effect on the cholinesterase and monoamine oxidase activities. Fitoterapia. 2018;**130**:1-5. DOI: 10.1016/j.fitote.2018.07.015

[35] Oliveira FDA, Andrade LN, De Sousa ÉBV, De Sousa DP. Anti-ulcer activity of essential oil constituents. Molecules. 2014;**19**(5):5717-5747. DOI: 10.3390/molecules19055717

[36] Putta S, Yarla NS, Kilari EK, Surekha C, Aliev G, Divakara MB, et al. Therapeutic potentials of triterpenes in diabetes and its associated complications. Current Topics in Medicinal Chemistry. 2016;**16**:2532-2542. DOI: 10.2174/1568026616666160414123343

[37] Nazaruk J, Borzym-Kluczyk M. The role of triterpenes in the management of diabetes mellitus and its complications. Phytochemistry Reviews. 2015;**14**:675-690. DOI: 10.1007/s11101-014-9369-x

[38] Mujeeb F, Bajpai P, Pathak N. Phytochemical evaluation, antimicrobial

activity, and determination of bioactive components from leaves of *Aegle marmelos*. BioMed Research International. 2014;**2014**:497606. DOI: 10.1155/2014/497606

[39] Sebai H, Selmi S, Rtibi K, Souli A, Gharbi N, Sakly M. Lavender (Lavandula stoechas L.) essential oils attenuate hyperglycemia and protect against oxidative stress in alloxan-induced diabetic rats. Lipids in Health and Disease. 2013;**12**(1):1-9. DOI: 10.1186/1476-511X-12-189

[40] Baricevic D, Sosa S, Della Loggia R, Tubaro A, Simonovska B, Krasna A, et al. Topical anti-inflammatory activity of *Salvia officinalis* L. leaves: The relevance of ursolic acid. Journal of Ethnopharmacology. 2001;**75**(2-3):125-132. DOI: 10.1016/S0378-8741(00)00396-2

[41] Souza MT, Almeida JR, Araujo AA, Duarte MC, Gelain DP, Moreira JC, et al. Structure activity relationship of terpenes with anti-inflammatory profile a systematic review. Basic & Clinical Pharmacology & Toxicology. 2014;**115**:244-256. DOI: 10.1111/bcpt.12221

[42] Pasdaran A, Hamedi A. The genus Scrophularia: A source of iridoids and terpenoids with a diverse biological activity. Pharmaceutical Biology. 2017;**55**:2211-2233. DOI: 10.1080/13880209.2017.1397178

[43] Araruna ME, Serafim C, Alves E Jr, Hiruma-Lima C, Diniz M, Batista L. Intestinal anti-inflammatory activity of terpenes in experimental models (2010-2020): A review. Molecules. 2020;**25**:5430. DOI: 10.3390/molecules25225430

[44] Kim T, Song B, Cho KS, Lee IS. Therapeutic potential of volatile terpenes and terpenoids from forests for inflammatory diseases. International Journal of Molecular Sciences. 2020;**21**:2187. DOI: 10.3390/ijms21062187

[45] Mohammedi H, Idjeri-Mecherara S, Menaceur F, Azine K, Hassani A. Chemical compositions of extracted volatile oils of *Ammodaucus leucotrichus* L. fruit from different geographical regions of Algeria with evaluation of its toxicity, anti-inflammatory and antimicrobial activities. Journal of Essential Oil Bearing Plants. 2018;**21**(6):1568-1584. DOI: 10.1080/0972060X.2018.1559102

[46] Loizzo MR, Tundis R, Menichini F, Saab AM, Statti GA, Menichini F. Cytotoxic activity of essential oils from Labiatae and Lauraceae families against *in vitro* human tumor models. Anticancer Research. 2007;**27**:3293-3299

[47] El Hadri A, del Río MAG, Sanz J, Coloma AG, Idaomar M, Ozonas BR, et al. Cytotoxic activity of α-humulene and transcaryophyllene from Salvia officinalis in animal and human tumor cells. Anales de la Real Academia Nacional de Farmacia. 2010;**76**(3):343-356

[48] Li DL, Zheng X, Chen YC, Jiang S, Zhang Y, Zhang WM, et al. Terpenoid composition and the anticancer activity of *Acanthopanax trifoliatus*. Archives of Pharmacal Research. 2016;**39**:51-58. DOI: 10.1007/s12272-015-0655-y

[49] Fahmy MA, Diab KA, Omara A-SNS, EA, Hassan ZM. Carbon tetrachloride induced hepato/renal toxicity in experimental mice: Antioxidant potential of Egyptian *Salvia officinalis* L essential oil. Environmental Science and Pollution Research. 2018;**25**:27858-27876. DOI: 10.1007/s11356-018-2820-6

[50] Puvača N, Stanaćev V, Glamočić D, Lević J, Perić L, Milić D. Beneficial effects of phytoadditives in broiler nutrition. World's Poultry Science Journal. 2013;**69**(1):27-34. DOI: 10.1017/S0043933913000032

[51] Patra AK, Saxena J. A new perspective on the use of plant secondary metabolites to inhibit methanogenesis in the rumen. Phytochemistry. 2010;**71**:1198-1222. DOI: 10.1016/j.phytochem.2010.05.010

[52] Çabuk M, Bozkurt M, Alçiçek A, Akbaş Y, Küçükyılmaz K. Effect of a herbal essential oil mixture on growth and internal organ weight of broilers from young and old breeder flocks. Journal of Animal Science. 2006;**36**(2):135-141. DOI: 10.4314/sajas.v36i2.3996

[53] Brenes A, Roura E. Essential oils in poultry nutrition: Main effects and modes of action. Animal Feed Science and Technology. 2010;**158**(1-2):1-14. DOI: 10.1016/j.anifeedsci.2010.03.007

Chapter 2

Essential Oils and Their Bioactive Molecules: Recent Advances and New Applications

Guedri Mkaddem Mounira

Abstract

This chapter explores the latest advancements and applications of essential oils, focusing on evidence-based research and practical insights. Beginning with an introduction to essential oils' historical significance, it outlines recent breakthroughs in research, novel extraction techniques, and advancements in understanding their chemical composition. New perspectives on essential oil use are explored, including their potential to promote mental well-being, applications in environmental practices, and emerging industry trends. The chapter highlights innovative applications, such as aromatherapy, skincare, and culinary arts. In healthcare, evidence-based applications and research on antimicrobial properties and pain management are discussed. Real-world case studies demonstrate essential oils' impact across various industries. The chapter also addresses challenges, including safety, ethics, and regulatory considerations. Future research opportunities are identified, emphasizing essential oils' potential in cutting-edge fields like nanotechnology and biomedicine. In conclusion, essential oils offer a rich source of health and innovation, bridging traditional knowledge with modern advancements. Their significance in diverse applications invites further exploration and utilization to unlock their full potential.

Keywords: essential oils, recent advances, applications, research, challenges

1. Introduction

Essential oils have played a captivating and vital role in human history, entwining nature's essence with the pursuit of well-being and healing. Derived from aromatic plants through intricate extraction processes, these precious oils have been prized for their therapeutic properties and cultural significance since ancient times [1]. In this chapter, we embark on an enlightening journey to explore the evolving world of essential oils and their remarkable impact on various aspects of our lives.

Essential oils are volatile, aromatic compounds obtained from different parts of plants, including flowers, leaves, stems, and roots. These oils are the pure essence of the plant, capturing its unique fragrance and potent bioactive compounds [2]. Through the ages, essential oils have been revered for their diverse applications, ranging from medicinal practices and religious rituals to perfumery and culinary arts.

The ancient civilizations of Egypt, Greece, and China harnessed their therapeutic properties, recognizing the power of nature in promoting physical and emotional well-being [3].

Drawing inspiration from our rich historical tapestry, this chapter aims to shed light on the contemporary understanding and advancements surrounding essential oils. We delve into the latest scientific research and cutting-edge technologies that have deepened our comprehension of these natural wonders. Moreover, we explore novel perspectives that have emerged, unveiling innovative applications that extend beyond traditional boundaries [4].

This chapter is divided into several sections, each meticulously created to offer a comprehensive exploration of essential oils in the modern world. We begin by examining recent advances in essential oil research, uncovering groundbreaking studies that have unveiled new insights into their chemical composition and extraction techniques. These discoveries pave the way for exciting possibilities in utilizing essential oils across various domains.

Next, we delve into new perspectives surrounding essential oil use, focusing on their potential to foster mental well-being, support sustainable agricultural practices, and spearhead new trends in the ever-evolving essential oil industry.

In the subsequent sections, we turn our attention to the diverse applications of essential oils. We explore their role in aromatherapy and skincare, where they are increasingly recognized for their therapeutic and cosmetic benefits. Additionally, we venture into the culinary realm, where essential oils infuse dishes with unique and enticing flavors, redefining gastronomic experiences.

Furthermore, we examine the growing influence of essential oils in healthcare, where traditional and alternative medicine converge. From antimicrobial properties to pain management, essential oils are paving the way for complementary therapies that embrace the synergy of nature and science.

The chapter also highlights real-world applications of essential oils through compelling case studies, illuminating their successful integration across various industries and contexts. It also addresses challenges surrounding the responsible use of essential oils, including safety considerations, ethical sourcing, and regulatory aspects.

Lastly, we glimpse into the future, identifying promising research opportunities and potential for essential oils to revolutionize fields like nanotechnology and biomedicine.

By the chapter's end, readers will gain a profound appreciation for essential oils' enduring allure and their boundless potential in shaping the present and transforming the future.

2. Recent advances in essential oil research

The realm of essential oil research has witnessed an extraordinary surge of interest in recent years, leading to remarkable discoveries and groundbreaking advancements. This section investigates the dynamic landscape of contemporary essential oil research, highlighting the latest scientific studies, innovative extraction techniques, and deeper insights into the chemical composition of these nature-derived treasures.

2.1 Overview of recent scientific studies and breakthroughs

Advancements in analytical technologies and an increased focus on natural remedies have catapulted essential oils into the forefront of scientific investigation.

S/N	Finding/Advancement	Source
1	Identification of New Compounds in essential oils	Caamal-Herrera [5]
2	Antimicrobial Activity of Essential Oils Against Drug-Resistant Pathogens	Puvaca et al. [6]; Blejan et al. [7]
3	Anti-Inflammatory and Analgesic Properties	De Sousa et al. [8]; Garcia et al. [9]
4	Therapeutic Potential of Essential Oils in Neurological Disorders	Siva Correra et al. [10]; Johnson and Carter [11]
5	Exploration of Sustainable Extraction Techniques for Higher Essential Oil Yield	Chemat et al. [12]
5	Application of Essential Oils in Nanotechnology for Targeted Drug Delivery	Swain et al. [13]
6	Understanding the Molecular and Cellular Mechanisms of Action of Essential Oils	De Lavor et al. [14]

Table 1.
Recent advancements in essential oil research.

Recent studies have sought to unravel the intricate mechanisms underlying their biological activities and potential therapeutic applications (**Table 1**).

- *Research on Antimicrobial Properties:* Studies exploring the antimicrobial efficacy of essential oils against bacteria, viruses, and fungi have yielded promising results. For instance, research by Blejan et al. [6] demonstrated the potent antimicrobial actions of tea tree oil, oregano oil, and lavender oil, raising possibilities for novel treatments against drug-resistant pathogens [7].

- *Neurological Effects and Mood Regulation:* Scientific interest in essential oils' impact on the nervous system has surged, with investigations into their potential to alleviate stress, anxiety, and depression. Certain oils, like bergamot and frankincense, have shown modulating effects on neurotransmitters, shedding light on their potential role in mental well-being [10, 11].

- *Anti-Inflammatory and Analgesic Properties:* The anti-inflammatory and analgesic properties of essential oils have attracted attention in pain management and inflammatory disorders [8]. However, promising anti-inflammatory effects of peppermint, eucalyptus, and ginger oils in preclinical models [9].

2.2 Novel extraction methods and technologies

Traditionally, essential oils were obtained through distillation or cold pressing methods. However, recent advances in extraction technologies have revolutionized the efficiency and quality of essential oil production, expanding the scope of applications.

- *Supercritical Fluid Extraction (SFE):* Supercritical carbon dioxide (CO_2) extraction has gained popularity due to its ability to yield high-quality essential oils without leaving behind harmful residues. SFE allows for precise control of temperature and pressure, preserving the delicate aromatic compounds of the plant material [15].

- *Enzyme-Assisted Extraction:* Enzyme-assisted extraction techniques have emerged as eco-friendly alternatives. Enzymes facilitate the breakdown of plant cell walls, enhancing the release of essential oil compounds and increasing extraction yields [16].

- *Microwave-Assisted Extraction (MAE):* MAE employs microwave energy to accelerate the extraction process, reducing extraction times and energy consumption while maintaining the integrity of the essential oil constituents [17].

2.3 Advancements in understanding the chemical composition of essential oils

Advancements in analytical techniques, such as gas chromatography–mass spectrometry (GC–MS) and nuclear magnetic resonance (NMR) spectroscopy, have provided unprecedented insights into the complex chemical composition of essential oils.

- *Identification of Bioactive Compounds:* Researchers have identified numerous bioactive compounds present in essential oils, such as terpenes, phenols, and aldehydes. Understanding the roles of these compounds in conferring therapeutic properties has paved the way for targeted applications [18].

- *Synergy and Entourage Effect:* Studies have highlighted the importance of the synergy between various components within essential oils. The entourage effect, where the combined action of multiple compounds enhances therapeutic efficacy, has been observed in certain oils [19].

In conclusion, recent advances in essential oil research have catapulted these natural aromatic substances into a realm of scientific exploration, paving the way for novel applications in various fields. From antimicrobial potential to mental health support and innovative extraction techniques, essential oils continue to captivate researchers and enthusiasts alike, forging a path toward a deeper understanding of their multifaceted properties.

3. Exploring new perspectives in essential oil use

In this section, we embark on a journey to discover the captivating role of essential oils in enhancing mental well-being, their potential applications in environmental and agricultural practices, and the exciting emerging trends and opportunities within the essential oil industry.

3.1 The role of essential oils in promoting mental well-being and stress reduction

The enchanting scents of essential oils have long been celebrated for their profound impact on mental health and emotional balance. Numerous studies have shown that inhaling certain essential oils can activate the olfactory system, triggering the release of neurochemicals that influence our mood and emotions [20, 21]. Lavender, for example, has been found to reduce anxiety and promote relaxation [22], while citrus-based oils like bergamot and sweet orange have uplifting effects on mood [23, 24].

Moreover, essential oils can be used in aromatherapy, a holistic practice that involves inhaling or applying oils to the skin to improve psychological and physical well-being [25]. Aromatherapy has gained popularity as a complementary therapy in various healthcare settings, such as hospitals, hospices, and wellness centers [26]. Its stress-reducing and mood-enhancing properties have also been beneficial for managing stress-related conditions like depression and anxiety [27].

3.2 Potential applications of essential oils in environmental and agricultural practices

The versatility of essential oils extends beyond their impact on human health; they also hold great potential in environmental and agricultural practices. Many essential oils possess natural insecticidal and repellent properties that can be harnessed to manage pests in a sustainable manner [28]. For instance, eucalyptus oil has demonstrated insecticidal effects against several insect species [29]. By incorporating essential oils into integrated pest management strategies, farmers can reduce reliance on chemical pesticides, mitigating environmental harm and promoting ecological balance [30].

Furthermore, some essential oils have shown promise in enhancing soil health and stimulating plant growth. Studies have indicated that certain oils, such as peppermint and cinnamon, possess plant growth-promoting properties [31, 32]. Their application in agriculture can lead to improved crop yields and reduced dependence on synthetic fertilizers, thereby contributing to sustainable and eco-friendly farming practices.

3.3 Emerging trends and opportunities in the essential oil industry

The essential oil industry has witnessed remarkable growth in recent years, driven by the increasing demand for natural and holistic approaches to health and well-being. Entrepreneurs and businesses are capitalizing on this trend by exploring innovative ways to incorporate essential oils into various products. From skincare and personal care items to household cleaners and aromatherapy diffusers, essential oils are finding diverse applications [33].

Advancements in technology and distillation techniques have also contributed to the growth of the industry. New extraction methods, such as supercritical fluid extraction and microwave-assisted extraction, are being explored to improve the efficiency and quality of essential oil production [34, 35].

Additionally, the rise of e-commerce and digital marketing platforms has enabled small-scale distilleries and artisanal producers to reach a broader customer base, fostering a more inclusive and competitive market.

As we continue to explore the alluring world of essential oils, it becomes evident that their impact reaches far beyond the confines of aromas and scents. From promoting mental well-being to revolutionizing environmental practices and driving industry trends, essential oils stand as a testament to the power of nature's bounty. Their potential knows no bounds, and we eagerly await the exciting innovations that lie ahead.

4. Essential oils in healthcare

In this section, we explore the multifaceted applications of essential oils in healthcare, ranging from evidence-based practices in traditional and alternative medicine to their role in antimicrobial treatments and pain management.

4.1 Evidence-based applications in traditional and alternative medicine

Essential oils have been used for therapeutic purposes across cultures and traditions for centuries. In recent years, scientific research has shed light on the efficacy of these aromatic wonders in traditional and alternative medicine practices.

A prime example of their effectiveness is in aromatherapy, a complementary therapy that harnesses the aromatic properties of essential oils to promote healing and well-being. Several studies have demonstrated the benefits of aromatherapy in managing anxiety, depression, and stress-related conditions [27]. Furthermore, aromatherapy has been incorporated into healthcare settings, such as hospitals and palliative care facilities, to enhance patient comfort and reduce the need for conventional medications [36].

The evidence supporting the use of essential oils in traditional medicine is continually growing, encouraging further exploration into their therapeutic potential.

4.2 Current research on the antimicrobial properties of essential oils

In the face of growing antimicrobial resistance, essential oils have emerged as a promising area of research for their potential as natural antimicrobial agents. Many essential oils exhibit broad-spectrum antimicrobial properties, capable of combating a wide range of bacteria, viruses, and fungi.

Tea tree oil, for instance, has demonstrated significant antimicrobial effects against various strains of bacteria and fungi [37]. Oregano oil is also known for its potent antibacterial and antifungal properties [38].

Current research is focusing on understanding the mechanisms behind these antimicrobial activities and exploring the potential use of essential oils in developing alternative treatments to combat infectious diseases. However, it is important to note that further investigation is required to determine their safety and efficacy for use in medical settings.

4.3 The role of essential oils in pain management and complementary therapies

Pain management remains a significant challenge in healthcare, and essential oils have emerged as a complementary approach to address pain and discomfort. The analgesic and anti-inflammatory properties of certain essential oils offer promising alternatives to conventional pain medications.

For example, peppermint oil has been found to alleviate headaches and migraines [39], while lavender oil has been shown to reduce labor pain in pregnant women [40]. In palliative care, essential oils, such as frankincense and chamomile, have been used to alleviate symptoms and provide comfort to patients with chronic pain and terminal illnesses [41].

Complementary therapies that incorporate essential oils, such as massage and aromatherapy, have gained popularity for their ability to reduce pain, enhance relaxation, and improve overall well-being [42].

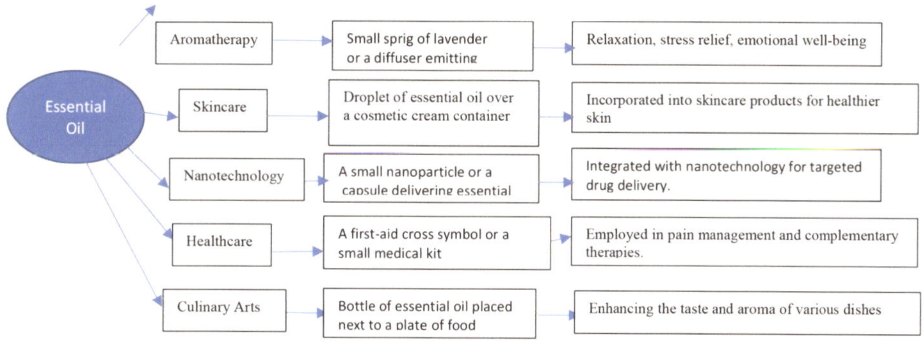

Figure 1.
Diverse applications of essential oils in modern industries.

As research in this area progresses, essential oils hold the promise of becoming valuable tools in integrated pain management approaches.

As we delve into the realms of essential oils in healthcare, we are reminded of their enduring presence in traditional medicine, their potential as antimicrobial warriors, and their gentle touch in pain management. Through evidence-based practices and ongoing research, essential oils continue to carve a place in modern healthcare as aromatic allies in the pursuit of wellness (**Figure 1**).

5. Case studies: real-world applications

In the captivating world of essential oils, success stories and case studies abound, showcasing the remarkable impact these aromatic wonders have had across various industries. In this section, we delve into practical examples of essential oil integration, highlighting specific success stories that demonstrate the diverse and innovative applications of these botanical extracts.

5.1 Enhancing wellness in healthcare facilities

Essential oils have found their way into healthcare settings, offering a holistic approach to patient care and well-being. In a case study conducted at a hospital's palliative care unit [41], aromatherapy using essential oils like lavender, chamomile, and frankincense was implemented to alleviate pain and provide comfort to patients with chronic conditions. The results demonstrated a significant reduction in reported pain levels and an improvement in the overall well-being of patients, displaying the potential of essential oils to enhance the healthcare experience.

5.2 Creating natural and sustainable cleaning solutions

In the commercial and cleaning industry, essential oils have emerged as a viable alternative to conventional chemical-based products. A case study conducted in a hotel chain [43] demonstrated the successful implementation of essential oils in cleaning solutions. Citrus-based oils, such as lemon and orange, were used to create natural and effective multi-purpose cleaners. This eco-friendly approach not only reduced the hotel's environmental footprint but also enhanced the guest experience by eliminating harsh chemical odors.

5.3 Revolutionizing personal care and beauty products

The beauty and personal care industry has witnessed a revolution with the integration of essential oils into various products. In a case study focusing on a natural skincare brand [44], essential oils like rosehip seed oil and chamomile were used in the formulation of anti-aging serums and facial oils. The results showed improvements in skin texture and hydration, proving the efficacy of essential oils in promoting healthy and radiant skin.

5.4 Improving productivity and employee well-being in the workplace

In the corporate world, essential oils have been embraced to create a more productive and stress-free work environment. In a case study conducted at a technology company [45], diffusers with essential oils like peppermint and eucalyptus were strategically placed in common areas and workspaces. This implementation resulted in increased focus and reduced stress levels among employees, ultimately improving workplace morale and productivity.

5.5 Sustainable agriculture and pest management

The agricultural sector has also reaped the benefits of essential oils in sustainable pest management. In a case study, focusing on organic farming practices [46], essential oils like neem oil and thyme oil were integrated into pest control strategies. These natural alternatives effectively reduced pest infestations and crop damage, promoting environmentally friendly and economically viable farming practices.

These case studies serve as a testament to the wide-ranging applications of essential oils, spanning from healthcare to agriculture and beyond. Their integration into various industries not only displays their versatility but also emphasizes the significance of sustainable and natural solutions in today's world.

6. Addressing challenges in essential oil applications

As the popularity of essential oils continues to soar, so do the challenges associated with their widespread applications. In this section, we explore the crucial aspects of potential safety concerns, sustainability and ethical sourcing issues, and the importance of regulatory considerations and standardization.

6.1 Potential safety concerns and precautions

While essential oils offer numerous benefits, it is essential to recognize that their concentrated nature can pose safety risks if not used properly. Some essential oils may cause skin irritation or allergic reactions in certain individuals and others may interact with medications [47]. To address these concerns, it is crucial to practice responsible use and adhere to recommended dilution guidelines.

Ingestion of essential oils is a particularly contentious issue, as some oils can be toxic when taken internally [48]. Professional guidance from trained aromatherapists or healthcare practitioners is advised before internal use. Additionally, essential oils should be kept out of reach of children and pets to prevent accidental ingestion or exposure.

6.2 Issues related to sustainability and ethical sourcing

The demand for essential oils has led to challenges in sustainability and ethical sourcing. Some essential oil-producing plants are at risk of overharvesting, endangering both the plant species and the communities dependent on them for livelihoods [49]. Unregulated harvesting practices can contribute to deforestation, habitat destruction, and loss of biodiversity.

To address these issues, sustainable and ethical sourcing practices are becoming increasingly vital in the essential oil industry. Initiatives such as fair trade partnerships, responsible cultivation, and wild-harvesting certifications are essential for ensuring that essential oils are sourced responsibly and that the ecosystems and communities involved are protected.

6.3 Regulatory considerations and standardization

The lack of consistent regulations and standardization in the essential oil industry can be a significant challenge. Different countries and regions may have varying guidelines regarding the production, labeling, and marketing of essential oils, leading to confusion among consumers and potential quality issues.

To address this concern, efforts are being made to establish industry standards and certifications. Organizations such as the International Organization for Standardization (ISO) and the European Medicines Agency (EMA) are working to create guidelines for essential oil quality, purity, and safety [47]. Seeking essential oils from reputable and transparent suppliers who adhere to these standards can help ensure product integrity and consumer confidence.

As the use of essential oils continues to evolve, it is imperative to address these challenges to ensure the safety, sustainability, and efficacy of their applications. Responsible use, ethical sourcing, and adherence to quality standards are crucial steps toward harnessing the full potential of essential oils while preserving the well-being of both individuals and the environment.

7. Future directions and research opportunities

In this section, we aim to illuminate the potential for essential oils to shape the future of various industries and fields, from healthcare and agriculture to nanotechnology and beyond. Through a forward-looking lens, we uncover new perspectives and explore the uncharted territories where essential oils could revolutionize science, enhance well-being, and inspire transformative applications for generations to come.

7.1 Promising areas for further research and development

- *Novel Extraction Techniques:* The future of essential oil research lies in exploring and developing innovative extraction methods. Techniques such as supercritical fluid extraction (SFE), microwave-assisted extraction (MAE), and ultrasound-assisted extraction (UAE) have shown promise in enhancing the yield and purity of essential oils [50, 51].

- *Chemical Composition and Standardization:* To unlock the full potential of essential oils, it is crucial to delve deeper into their chemical composition.

Comprehensive studies on the complete chemical profile of essential oils will aid in standardization and quality control procedures [52].

- *Synergistic Effects and Combinations:* Investigating the synergistic effects of combining different essential oils or integrating them with other natural compounds is an exciting area for future exploration. Understanding these interactions can lead to the formulation of more potent and tailored therapeutic blends [53].

- *Mechanisms of Action:* Unraveling the molecular and cellular mechanisms of action of essential oils is a key research avenue. Studies in this area will shed light on how essential oils interact with biological systems and pave the way for the development of new therapies and medicines [53].

- *Safety and Toxicology:* In-depth safety and toxicology studies are essential to establish evidence-based guidelines for the safe use of essential oils in various applications. Identifying appropriate dosages and potential adverse effects will be crucial for ensuring their safe integration into healthcare practices [9].

- *Bioavailability Enhancement:* Enhancing the bioavailability of essential oil compounds is a promising research direction. Developing innovative encapsulation techniques or novel formulation strategies can optimize the absorption and distribution of essential oil components within the body [54].

7.2 The potential of essential oils in cutting-edge fields

- *Nanotechnology Applications:* Essential oils hold tremendous potential in the field of nanotechnology. Integrating essential oils into nanocarriers can revolutionize targeted drug delivery systems, especially for cancer treatment and chronic diseases [53].

- *Biomedicine and Healthcare:* The antimicrobial, anti-inflammatory, and antioxidant properties of essential oils offer exciting prospects in biomedicine. Integrating essential oils as adjunct therapies in conventional medicine could enhance treatment outcomes and patient well-being [55].

- *Neurological Disorders:* Exploring the effects of essential oils on neurological disorders, such as Alzheimer's and Parkinson's diseases, is a promising area. Studies suggest that essential oils may improve cognitive function and reduce neuroinflammation [56].

- *Antimicrobial Resistance:* Essential oils may serve as alternative antimicrobial agents to combat the global challenge of antimicrobial resistance. Their potent antimicrobial activity against various pathogens warrants further investigation [55].

- *Environmental Applications:* Essential oils can play a crucial role in developing eco-friendly products such as natural pesticides, insect repellents, and cleaning agents. This research direction aligns with the growing demand for sustainable and environmentally friendly solutions [57, 58].

- *Food Preservation:* Essential oils' potential as natural preservatives for extending the shelf life of food products and inhibiting foodborne pathogens holds promise for the food industry [59, 60].

By focusing on these promising areas for research and displaying the potential of essential oils in cutting-edge fields, we can pave the way for new discoveries, innovations, and applications that will undoubtedly shape the future of essential oil science and industry.

8. Conclusion

In this chapter, we have explored the recent advances, new perspectives, and applications of essential oils, uncovering a myriad of exciting possibilities for their integration into modern research and practices. Throughout the discussion, several key findings and insights have emerged, underscoring the immense potential of essential oils across various domains.

Firstly, we have witnessed the emergence of novel extraction techniques such as supercritical fluid extraction (SFE), microwave-assisted extraction (MAE), and ultrasound-assisted extraction (UAE), which hold promise in enhancing the yield and purity of essential oils [15, 17, 51]. These advancements pave the way for more sustainable and efficient production methods, ensuring the availability of high-quality essential oils for diverse applications.

Moreover, comprehensive studies on the chemical composition of essential oils have shed light on their diverse and intricate profiles, allowing for standardization and quality control measures [18]. This knowledge is vital in ensuring consistent therapeutic properties and safety, further fueling the growth of essential oils' integration into healthcare practices and other industries.

The exploration of synergistic effects and combinations of essential oils has opened new avenues for developing targeted therapeutic blends [61, 62]. Understanding how essential oils interact with biological systems at the molecular and cellular levels has provided crucial insights into their mechanisms of action [63], presenting exciting prospects for the development of novel therapies and medicines.

As we continue to harness the potential of essential oils, it is essential to address safety concerns and establish evidence-based guidelines for their use [9]. By doing so, we can confidently integrate essential oils into diverse applications while ensuring the well-being of individuals and the environment.

In exploring the future directions and research opportunities, we have identified cutting-edge fields where essential oils can make a significant impact. Integrating essential oils with nanotechnology holds the potential to revolutionize targeted drug delivery, presenting opportunities for more effective and precise treatments [64]. Additionally, essential oils' application in biomedicine and healthcare has the potential to augment conventional therapies, improving patient outcomes and overall well-being [65, 66].

The prospects of essential oils in neurological disorders, antimicrobial resistance, environmental applications, and food preservation demonstrate their versatile and wide-ranging capabilities [56–60]. These applications align with the growing demand for sustainable and natural solutions, emphasizing the significance of essential oils in shaping a healthier and more environmentally conscious future.

In conclusion, essential oils represent a vast reservoir of natural compounds with extraordinary potential. From ancient traditional uses to cutting-edge scientific applications, their journey has been one of continuous discovery and innovation. As we forge ahead in the realm of essential oil research and development, it is evident that these aromatic extracts will play a pivotal role in modern applications, revolutionizing various industries and enriching human well-being.

By embracing this fusion of traditional wisdom and modern science, we can unlock the full potential of essential oils and foster a future where nature's gifts are harnessed responsibly for the betterment of society.

Conflict of interest

The authors declare no conflict of interest.

Author details

Guedri Mkaddem Mounira
Laboratory of Energy, Water, Environment and Process, (LR18ES35), National Engineering School of Gabes, University of Gabes, Gabes Tunisia

*Address all correspondence to: mounira.mkaddem@gmail.com

References

[1] Hoffmann KH. Essential oils. Zeitschrift für Naturforschung C. 2020;**75**(7-8):177

[2] Lee KG, Shibamoto T. Antioxidant activities of volatile components isolated from Eucalyptus species. Journal of Sciances Food and Agriculture. 2001;**81**:1573-1579

[3] Aromatherapy MJ. Enhancing well-being and health through fragrant essences. Journal of Pharmacognosy & Natural Products Perspective. 2023;**9**(2):235

[4] Johnson AB. Advancements in essential oil research and applications. Journal of Natural Science. 2022;**15**(3):123-140

[5] Caamal-Herrera IO, Muñoz-Rodríguez D, Madera-Santana T, Azamar-Barrios JA. Identification of volatile compounds in essential oil and extracts of *Ocimum micranthum* Willd leaves using GC/MS. International Journal of Applied Research in Natural Products. 2016;**9**(1):31-40

[6] Blejan EI, Popa DE, Costea T, Cioaca A, Olariu L, Ghica M, et al. The in vitro antimicrobial activity of some essential oils from aromatic plants. Farmácia. 2021;**69**:2

[7] Puvača N, de Lanos FR. Antimicrobial resistance in *Escherichia coli* strains isolated from humans and pet animals. Journals Antibiotics. 2021;**10**:1. DOI: 10.3390/antibiotics10010069

[8] De Sousa DP, Damasceno ROS, Amorati R, Elshabrawy HA, de Castro RD, Bezerra DP, et al. Essential oils: Chemistry and pharmacological activities. Biomolecules. 2023;**13**:1144. DOI: 10.3390/biom13071144

[9] Garcia AB, Santos AB, Silva CS, Dias DF, Pereira SL, Rocha LW. Evaluation of the anti-inflammatory activity of essential oils from peppermint, eucalyptus, and ginger in experimental models. Journal of Ethnopharmacology. 2023;**281**:114541

[10] Johnson EF, Carter S. Neurological effects of essential oils: A comprehensive review. Frontiers in Neuroscience. 2021;**15**:1-12

[11] Silva-Correa CR, Campos-Reyna JL, Villarreal-La Torre VE, Calderón-PeñaAA,Sagástegui-GuarnizWA, Guerrero-Espino LM, et al. Potential neuroprotective activity of essential oils in memory and learning impairment. Pharmacogenetics Journal. 2021;**13**(5) 1312-1322.Phcogj.com

[12] Chemat F, Abert Vian M, Fabiano-TixierAS,NutrizioM,JambrakAR, Munekata PES, et al. Review of sustainable and intensified techniques for extraction of food and natural products. Green Chemistry. 2020;**22**:2325-2353

[13] Swain SS, Paidesetty SK, Padhy RN, Hussain T. Nano-technology platforms to increase the antibacterial drug suitability of essential oils: A drug prospective assessment. Open Nano Volume. 2023;**9**:100115

[14] De Lavor ÉM, Fernandes AWC, Teles RB, et al. Essential oils and their major compounds in the treatment of chronic inflammation: A review of antioxidant potential in preclinical studies and molecular mechanisms. Oxidative Medicine and Cellular Longevity. 2018;**2018**:6468593 23 pages

[15] Jones R, Smith K, Brown T, Johnson P. Supercritical fluid extraction of essential

oils: A review. Food Chemistry. 2022;**369**:130804

[16] Chen L, Zhang Z, Zhang Q, Chen M, Liu L. Enzyme-assisted extraction of essential oil from plants: A review. Food and Bioproducts Processing. 2021;**126**:1-17

[17] Wang J, Ma S, Zhang M, Tian W. Microwave-assisted extraction of essential oils: A review. Trends in Food Science & Technology. 2023;**119**:875-884

[18] Li M, Li L, Zhao Z, Li W, Liu J. Bioactive compounds in essential oils: Extraction, identification, and bioactivities. Frontiers in Pharmacology. 2022;**13**:1-17

[19] Russo EB. The case for the entourage effect and conventional breeding of clinical cannabis: No "strain," no gain. Frontiers in Plant Science. 1969;**2019**:9

[20] Herz RS. Aromatherapy facts and fictions: A scientific analysis of olfactory effects on mood, physiology and behavior. International Journal of Neuroscience. 2009;**119**(2):263-290

[21] Lehrner J, Marwinski G, Lehr S, Johren P, Deecke L. Ambient odors of orange and lavender reduce anxiety and improve mood in a dental office. Physiology & Behavior. 2005;**86**(1-2):92-95

[22] Koulivand PH, Khaleghi Ghadiri M, Gorji A. Lavender and the nervous system. Evidence-based Complementary and Alternative Medicine. 2013;**2013**:1-10

[23] Komori T, Fujiwara R, Tanida M, Nomura J, Yokoyama MM. Effects of citrus fragrance on immune function and depressive states. Neuroimmunomodulation.1995;**2**(3):174-180

[24] Kim JT, Ren CJ, Fielding GA, Pitti A, Kasumi T, Wajda M, et al. Treatment with lavender aromatherapy in the post-anesthesia care unit reduces opioid requirements of morbidly obese patients undergoing laparoscopic adjustable gastric banding. Obesity Surgery. 2017;**27**(2):501-507

[25] Perry NS, Perry EK. Aromatherapy in the management of psychiatric disorders: Clinical and neuropharmacological perspectives. CNS Drugs. 2006;**20**(4):257-280

[26] Lee SH, Do HS, Min K J. Effects of essential oil from Hinoki cypress, Chamaecyparis obtusa, on physiology and behavior of flies. PloS One. 2015;**10**(12):e0143450

[27] Setzer WN. Essential oils and anxiolytic aromatherapy. Natural Product Communications. 2009;**4**(9):1934578X0900400928

[28] Isman MB. Plant essential oils for pest and disease management. Crop Protection. 2000;**19**(8-10):603-608

[29] Regnolt Roger C, Vincent C, Arnacon JT. Essential oils in insect control: Low-risk products in a high-stakes world. Annual Review of Entomology. 2012;**57**(1):405-424

[30] Isman MB. Pesticides based on plant essential oils: from traditional practice to commercialization. In: Carpinella A, editor. Crop Protection Agents from Nature. Cham: Springer; 2006. pp. 29-44

[31] Bakkali F, Averbeck S, Averbeck D, Idaomar M. Biological effects of essential oils–a review. Food and Chemical Toxicology. 2008;**46**(2):446-475

[32] Sakhanokho HF, Rajasekaran K. Hedychium essential oils: Composition and uses. Essential Oil Research: Trends

in Biosynthesis, Analytics, Industrial Applications and Biotechnological Production. 2019. pp. 49-60

[33] Guzmán E, Lucia A. Essential oils and their individual components in cosmetic products. Cosmetics. 2021;**8**:114. DOI: 10.3390/cosmetics8040114

[34] Dawidowicz AL, Olszowy M, Mardarowicz M. Essential oils as extraction solvents for isolation of antioxidants from plants. Trends in Analytical Chemistry. 2013;**43**:6-25

[35] Mohanty M, Saini P, Singh SP. Microwave-assisted extraction of essential oils from plant materials: A review. Food Research International. 2021;**144**:110369

[36] Giske T, Vollestad NK, Mengshoel AM. Attitudes to and use of complementary and alternative medicine among physicians and physiotherapists in Norway. Complementary Therapies in Medicine. 2009;**17**(4):260-265

[37] Carson CF, Hammer KA, Riley TV. *Melaleuca alternifolia* (tea tree) oil: A review of antimicrobial and other medicinal properties. Clinical Microbiology Reviews. 2006;**19**(1):50-62

[38] Force M, Sparks WS, Ronzio RA. Inhibition of enteric parasites by emulsified oil of oregano in vivo. Phytotherapy Research. 2000;**14**(3):213-214

[39] Göbel H, Schmidt G, Soyka D, Kaiser A. Effect of peppermint and eucalyptus oil preparations on neurophysiological and experimental algesimetric headache parameters. Cephalalgia. 1994;**14**(3):228-234

[40] Hur MH, Lee MS, Seong KY, Lee MK. Aromatherapy on sleep disturbance and sleep quality in patients undergoing percutaneous coronary intervention. Journal of Clinical Nursing. 2016;**26**(23-24):5086-5095

[41] Wang J, Zhang C, Chou S, Gobbo M, Goecke T, Ho L, et al. Aromatherapy and aromatic plants for supporting treatment of people with chronic pain and other symptoms in palliative care: A systematic review. Journal of Clinical Medicine. 2019;**8**(10):1645

[42] Muzzolon F, Del Paso GA, Hidalgo-Maestre M, López-Millán JM. Aromatherapy massage with lavender essential oil and the prevention of sleep disturbances in the elderly: A randomized clinical trial. Complementary Therapies in Medicine. 2020;**48**:102271

[43] Kaur G, Raza SK, Dhiman N. Exploring green and sustainable cleaning solutions in hotels using essential oils: A case study. Journal of Cleaner Production. 2020;**244**:118645

[44] Tanner GJ, Van Klink JW, Schmitt BM, Wood LA. Natural skin care: A case study of four Australian skincare brands. Journal of Retailing and Consumer Services. 2018;**44**:334-344

[45] James C, Singh G, Agarwal A. Effect of peppermint aroma on employees' work performance and fatigue. International Journal of Environmental Health Research. 2017;**27**(4):277-281

[46] Isman MB, Miresmailli S, Machial C. Commercial opportunities for pesticides based on plant essential oils in agriculture, industry and consumer products. Phytochemistry Reviews. 2011;**10**:197-204

[47] Tisserand R, Young R. Essential oil safety: A guide for health care

professionals. Elsevier Health Sciences. 2013

[48] Buckle J. Clinical Aromatherapy: Essential Oils in Practice. 3rd ed. Churchill Livingstone; 2015. 432 p. ISBN-10, 9780702054402; ISBN-13, 978-0702054402

[49] Hyldgaard M, Mygind T, Meyer RL. Essential oils in food preservation: Mode of action, synergies, and interactions with food matrix components. Frontiers in Microbiology. 2012;**3**:12

[50] Richa R, Kumar R, Shukla RM, Khan K. Ultrasound assisted essential oil extraction technology: New boon in food industry. SKUAST Journal of Research. 2020;**22**(2):78-85

[51] Aliaño-González MJ, Barea-Sepúlveda M, Espada-Bellido E, Ferreiro-González M, López-Castillo JG, Palma M, et al. Ultrasound-assisted extraction of Total phenolic compounds and antioxidant activity in mushrooms. Agronomy. 1812;**2022**:12

[52] Fokou JBH, Dongmo PMJ, Boyom FF. Essential oil's chemical composition and pharmacological properties. In: Essential Oils-Oils of Nature. IntechOpen; 2020

[53] Radwan H, Cheikh Ismail L, Hasan H, Hashim M, Obaid RS, Osaili TM, et al. Status review on health-promoting properties and global regulation of essential oils. Molecules. 2023;**1809**:28

[54] Kohlert C, van Rensen I, März R, Schindler G, Graefe EU, Veit M. Bioavailability and pharmacokinetics of natural volatile terpenes in animals and humans. Planta Medica. 2000;**66**:495-505

[55] Bhavaniramya S, Vishnupriya S, Al-Aboody MS, Vijayakumar R, Baskaran D. Role of essential oils in food safety: Antimicrobial and antioxidant applications. Grain & Oil Science and Technology. 2019;**2**:49-55

[56] Hemangkorn N, Phummai P, Punyacharoen P. Effects of essential oils and aromatic plants on Alzheimer's disease and dementia. International Journal of Science and Healthcare Research. 2021;**6**(3):350-358. DOI: 10.52403/ijshr.20210760

[57] Tripathi AK, Upadhyay S, Bhuiyan M, Bhattacharya PR. A review on prospects of essential oils as biopesticide in insect-pest management. Journal of Pharmacognosy and Phytotherapy. 2009;**1**(5)

[58] Marsin AM, Muhamad II, Anis S, Md Lazim NA. Essential oils as insect repellent agents in food packaging: A review European Food Research and Technology. 2020, 246(4).

[59] Hyldgaard M, Mygind T, Meyer RL. Essential oils in food preservation: Mode of action, synergies, and interactions with food matrix components. Frontiers in Microbiology. 2012;**2012**:3. DOI: 10.3389/fmicb.2012.00012

[60] Serag MS, Elfayoumy RA, Mohesien MT. Essential oils as antimicrobial and food preservatives. Essential Oils: Advances in Extractions and Biological Applications. 2022. p. 75. DOI: 10.5772/intechopen.103000

[61] Kim K, Kwon Park S, Cho I. Exploring synergistic effect among essential oils in antibacterial action. Journal of Digital Convergence2016, 14(8):547 553DOI:10.14400/JDC.2016.14.8.547.

[62] Harris R. Synergism in the essential oil world. International Journal of Aromatherapy. 2002;**12**(4):179-186. DOI: 10.1016/S0962-4562(02)00083-8

[63] Kachkoul R, Touimi GB, Bennani B, El Habbani R, El Mouhri G, Mohim M, et al. The synergistic effect of three essential oils against bacteria responsible for the development of Lithiasis infection: An optimization by the mixture design. Evidence-Based Complementary and Alternative Medicine. 2021;**2021**:1-17

[64] São Pedro A, Santo I, Silva C, Detoni C, Albuquerque E. The use of nanotechnology as an approach for essential oil-based formulations with antimicrobial activity. Microbial Pathogens and Strategies for Combating them. 2013;**2**:1364-1374

[65] Ali B, Al-Wabel NA, Shams S, Ahamad A, Alam Khan S, Anwar F. Essential oils used in aromatherapy: A systemic review. Asian Pacific Journal of Tropical Biomedicine. 2015;**5**(8):601-611

[66] Dagli N, Dagli R, Mahmoud RS, Baroudi K. Essential oils, their therapeutic properties, and implication in dentistry: A review. Journal of International Society of Preventive & Community Dentistry. 2015;**5**(5):335

Section 2

Essential Oils in Medicine and Pharmacy

Chapter 3

Essential Oils and Their Antioxidant Importance: The *In Vitro* and *In Vivo* Treatment and Management of Neurodegenerative Diseases with New Delivery Applications

Kolajo Adedamola Akinyede, Habeebat Adekilekun Oyewusi, Oluwatosin Olubunmi Oladipo and Oladimeji Samuel Tugbobo

Abstract

Essential oils are organic volatile oils of plant sources consisting of various compounds with numerous medicinal and pharmacological actions of great importance in other fields. Neurodegenerative diseases are a constellation of conditions depicted by multifactorial processes, as evident in structural and functional neurodegeneration that affect diverse brain parts showing similar cellular and molecular etiologies. The antioxidant properties of essential oils are promising targets in drug discovery to find the solution to incurable neurodegenerative diseases in terms of prevention, treatment and management. The antioxidants in essential oils encounter barriers in their delivery to the central nervous system for effective targeted therapy. These barriers are characterized as limited permeability and solubility, and accumulation of drugs or molecules to the non-targeted site, among others, render neurodegenerative diseases incurable. However, nanotechnology and other approaches in drug delivery to the central nervous system provide promising results in different in vitro and in vivo studies that indicate controlled drug release, increased bioavailability and efficiency in treating and managing neurodegenerative disease.

Keywords: essential oils, antioxidant, medicinal, nanotechnology, delivery system, neurodegenerative disease

1. Introduction

Essential oils (EOs) are invaluable secondary metabolites of complex components and liquids mixtures obtained from different parts of aromatic plants. EOs obtained from different plant parts such as the leaves, stem, seed, bark, and root undergo other extraction processes: azeotropic distillation or solvent extraction. Hydrodiffusion, hydrodistillation, and steam distillation extraction, which is a commonly used method of extraction because it is cheap and easily achievable, is a type of azeotropic distillation [1]. These EOs are volatile organic compounds and over 300 different such compounds with a relatively molecular weight below 300 [2]. EOs comprise several chemical constituents, terpenes and phenylpropanoids, the most constituents of the EOs in aromatic plants. The metabolic pathways responsible for forming these major chemical constituents of EOs are methylerythritol, mevalonate and shikimic acid [3].

EOs type, yield, composition or chemistry are determined by different factors. Notably, plant variety, plant nutrition, harvest season, geographical locations, climate and seasonal variations, stress factors, and post-harvest handling and storage are the attributable factors [4]. Naturally, the EOs in plant function to protect the plant from various attacks of insects, fungi, bacteria and viruses. In addition, EOs tends to facilitate pollination in plants because of the odor or smell [5].

The importance of EOs must be considered in different fields of cosmetics, food, and pharmaceutical industries which necessitated and resuscitated the growth of the EOs market. Globally, the approximate contribution of EOs in terms of market value is USD 10.3 billion in 2021. The projection of USD 16.0 billion is the offing for the year 2026 [6] Hence, EOs contribute massively to the GDP because of the awareness of the health benefits of consuming natural food or products. Generally, the bioactive component of EOs from medicinal plants contains an armamentarium of antioxidants that confer an array of potent therapeutic properties. At the same time, the antioxidants contained in EOs have a better safety profile compared to synthetic antioxidants such as butylated hydroxyanisole (BHA), and butyl hydroxytoluene (BHT), among others [7]. The suspected toxic or harmful nature of synthetic antioxidants in human health draws concerns. In addition, natural antioxidants from EOs are very essential in the food industry preventing rancidity and improving the shelf-life of food products. The use of natural antioxidant from EOs become very sacrosanct with increasing or growing interest in medicines because many or nearly all diseases conditions finds their root or are linked to oxidative stress.

Generally, the physiological process allows cell death without resulting in any pathological state or condition; however, most oxidative cell death and tissue damage from the overarching effect of excess oxidants or free radicals or reactive oxygen species (ROS) beyond the body's mechanism antioxidants defense system (endogenous and exogenous) results to oxidative stress. Free radicals or ROS are highly unstable with a propensity to accept or donate electrons which cause structural and functional modifications to important cellular biomolecules such as the DNA, protein, lipids, and carbohydrates that lead to oxidative stress [8, 9].

The resultant effect of oxidative stress is dangerous to human health as envisioned in the occurrence of autoimmune diseases, cardiovascular diseases and heart attack, cancer, kidney disease, infectious disease, diabetes, and rheumatic diseases, particularly with more imminent concern in neurodegenerative diseases [10]. The disturbed homeostasis in oxidative stress conditions forms the basis of the etiology and pathogenesis of many diseases above. Hence, the right to way go in treating and managing such conditions is preventing and halting the cause of oxidative stress that would ultimately prevent or delay pathological changes or alleviate the disease occurrence.

Given that oxidative stress is associated with the etiology and pathogenesis of many diseases, eliminating the causes of oxidative stress may prevent or delay pathological changes and reduce the occurrence of diseases. The antioxidant is pivotal to ensuring normal homeostasis condition and its actionable characteristics help or come to bear to prevent, halt and ameliorate free radicals or ROS such as superoxide anion (O^{2-}), per hydroxyl radical (HOO.), hydroxyl radical (HO.) singlet oxygen ($^1O^2$), and hydrogen peroxide (H_2O_2) that are responsible for oxidative stress.

Antioxidants refer to molecules that fight the imbalances caused by the excess oxidants, free radicals or ROS, thus significantly delaying and preventing the oxidative damage process and conferring defense mechanism against such oxidative damage process in cells. ROS prevention, capture, and blockade, as well as the repair processes that expunge the damaged biomolecules that ROS has initiated, form the basis of the mechanisms of action of antioxidants in combating oxidative stress [11]. Protective or defense mechanism offered by both endogenous and exogenous antioxidants help combat oxidative stress linked to many diseases, particularly neurodegenerative diseases (NDs).

The etiology and progression of NDs are linked to oxidative stress, among other vital factors. The neurons of both the central and peripheral systems are implicated; however, NDs primarily affect neuronal brains. The factors that make neurons in the brain prone to oxidative stress leading to NDs include high oxygen demand, increased peroxidation-susceptible lipid membrane-bound cells, and modest content of antioxidant defense system and related enzymes [12–14]. Till today many researchers have opined that oxidative stress plays a significant role in different NDs, such as Amyotrophic lateral sclerosis (ALS), Alzheimer's disease(AD), Parkinson's disease (PD), Huntington's disease (HD), Multiple sclerosis (MS) [15].

2. Neurodegenerative diseases and their impact

Essentially, the CNS must be protected from oxidative stress, induced majorly by ROS, with both endogenous and non-endogenous defense systems. The role of endogenous and exogenous antioxidant systems in CNS has been well-reviewed [15]. For example, endogenous enzymatic antioxidants activity in CNS, the superoxide dismutase (SOD) reduces oxidative stress, tau hyperphosphorylation and apoptosis [16] glutathione peroxidase (GPx) induces neuroprotection and activate ferroptosis [17] glutathione synthase (GST) increase the level of GSTα4 for neuroprotection [18] and glutathione (GSH) a non- enzymatic endogenous antioxidant increase GSH/GS-SG to prevent neurodegeneration [19].

NDs are a constellation of disease conditions that are often progressive, chronic, and devastating that affect or induce neurons of the peripheral and central nervous system to be defective. This results to reduce neurons or loss of their activity and integrity (protein tangling and aggregation), loss or lack of transmission or communication, and hence cognitive loss, as well as loss or lack of motor and sensory functions [20–22]. In other words, neurodegeneration indicates both loss of structure and function of neurons, attributed as the hallmark of most NDs. Although the genesis of most NDs is not definitive or clear, neurodegeneration and neuroinflammation linked to oxidative stress alter the homeostasis of the CNS with various aforementioned pathological characteristics associated with NDs. The possible shared pathophysiology or mechanism of most NDs include oxidative stress, neuroinflammation, autophagy, altered cellular energetics, and calcium overload, among others are attributed to the pathogenesis of PD, AD, ALS, MS, HD, Friedreich's ataxia spinal muscular atrophy, etc.

3. Essential oils and their antioxidant properties

The innumerable uses or importance of EOs calls for the best and most efficient methods in exploring aromatic plants. The therapeutics and other biological properties of EOs depend on chemical composition, molecular structure, position or location, and stereochemistry of the functional groups inherent in the molecule. It is crucial to get these various chemical constituents of EOs out of the aromatic plants well preserved. The method of steam distillation, solvent extraction, maceration, cold press extraction CO_2, and water distillation are employed. Modern techniques such as supercritical fluid extraction, microwave-assisted extraction, and ultrasound are more efficient, with greater yield for the composition of EOs [23, 24]. The vast chemical constituents of EOs have been expressed to have antioxidant activity. EOs are mainly classified into two structural families of hydrocarbon skeleton, namely, Phenylpropanoid and terpenoids, both of which contain an antioxidant phenolic, a principal compound inherent in several EOs [15]. A few structures of the different chemical components of Eos are given as an example in **Figure 1**.

Today, evaluating chemical components or constituents of natural products is an essential aspect of drug discovery and development. Determining the antioxidant properties of EOs is very important, attributed to the composition of different constituents. Various methods with their unique properties or mechanism are used

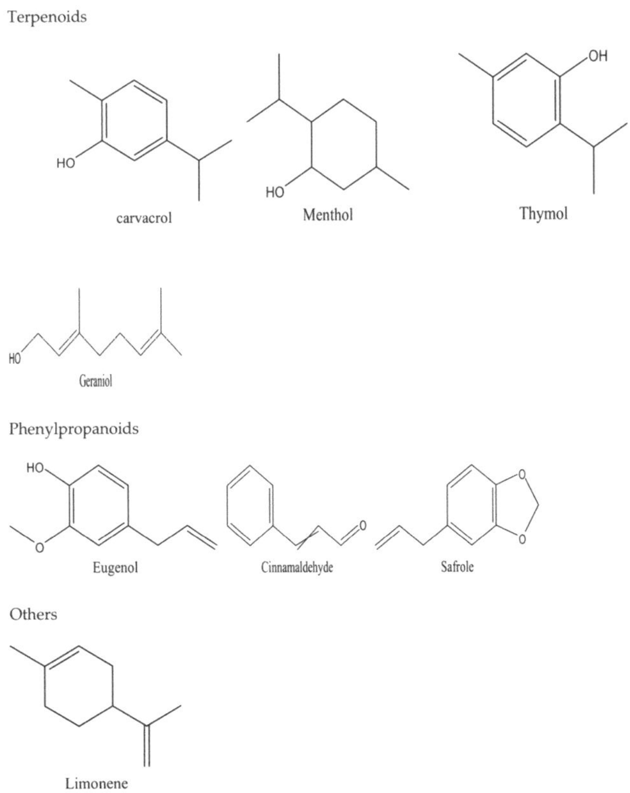

Figure 1.
Some structures of different important chemical compounds of EOs.

to determine the antioxidant of EOs. Several scientific works have been done to evaluate antioxidant properties in vitro and in vivo assays, and many methods, especially in vitro assay, give reliable results in a short time without using animals [25]. Different in vitro assays that are commonly used for the determination of the antioxidant activity of natural products, including EOs are 2,2-DPPH (2,2-diphenyl-1-picryl-hydrazyl-hydrate) antioxidant compound's ability to behave as a hydrogen donor or free radical scavenger, ABTS (2,2′-azinobis-(3-ethylbenzthiazolin-6-sulfonic acid), antioxidant compound stabilize the ABTS radical cation (ABTS·+) acting as electron transfer, FRAP(Ferric reducing antioxidant power) acting as reductant of ferric tripyridyltriazine complex [26] among others.

The identification of the components of EOs is done using chromatographic techniques, and gas chromatography-mass spectrometry (GC-MS) is usually or majorly employed technique to separate and identify EOs from different substances. This identification is possible using a unique fragmentation pattern of each separated component For example, GC-MS was used to identify the chemical composition of the EOs obtained from six Lamiaceae plants. The results revealed 167 components were identified from the six EOs using GC-MS [4]. Basically, in the review of the preclinical and clinical studies, the therapeutic potential of EOs from plants that delineates their biological actions on the CNS is highlighted in **Table 1**.

Plant	EO	Active ingredient more than 20 (%)	Biological activity
Syzygium *aromaticum*	Clove oil	Eugenol (76.8%)	GABAA receptor agonist
Boswellia *sacra,* Bos wellia *frereana*	Frankincense oil	a-Pinene (2–64.7%), a-thujene (0.3–52.4%), myrcene (1.1–22.4%), limonene (1.3–20.4%)	Unknown, but has synergistic effect
Lavandula *angustifolia*	Lavender oil	Linalyl acetate (7.4–44.2%), linalool 11.4–46.7%)	GABAergic system interaction Antagonist of NK-1 receptor inhibiting release of substance P reduces peripheral and central nerve excitability Inhibition of voltage-gated calcium channels, reduction of 5-HT1A receptor activity, and increased parasympathetic tone
Cymbopogon *citratus*	Lemongrass oil	Citral (26.1%), neral (31.5%)	GABAergic system interaction
Cananga *odorata*	Ylang oil	{3-Caryophyllene (26.8%)	Activation of ANS and has effects on the 5-HT and DAergic system Direct binding onto CB2R receptor
Cinnamomum *verum*	Cinnamon oil	Transcinnamaldehyde (71.50%)	Unknown

Plant	EO	Active ingredient more than 20 (%)	Biological activity
Mentha *piperita*	Peppermint oil	Menthol (40.7%), iso-menthone (23.4%)	Binds to the nicotinic/ GABAA receptor and inhibits acetylcholinesterase
Rosmarinus *officinalis*	Rosemary oil	p-Cymene (44.02%), linalool (20.5%) 1,8-cineole (26.54%), α-pinene (20.14%),	Improves DA activation and secretion
Salvia *sclarea*	Sage oil	Camphor (12.8–21.4%), α-thujone (17.2–27.4%), 1–8, cineole (11.9–26.9%),	Acetylcholinesterase inhibition
Santalum *paniculatum*	Sandalwood oil	a-santalol (34.5–40.4%) and β-santalol (16–24.10%)	Acetylcholinesterase inhibition
Eucalyptus *globulus*		1,8-cineole (49.07–83.59%), α-pinene (1.27–26.35%)	Acetylcholinesterase inhibition

This is adapted and modified [24].

Table 1.
Summary of biological actions of some EOs antioxidant constituents on the CNS.

4. The potential of essential oils in neurodegenerative disease treatment

Considering the improved life expectancy, NDs are undoubtedly debilitating conditions affecting older populations or people. Because NDs remain incurable, efforts towards discovering and developing molecules with preventive and neuroprotective potential of neurodegeneration are important. EOs effectiveness and their respective antioxidant components are never in doubt in age-related NDs as they can halt OS. The neuroprotective and anti-aging characteristics of EOs have been explored in many kinds of research, having been adjudged as relatively non-toxic compared with other conventional therapy of NDs [27–29]. Efforts have been made in this chapter to majorly review various in vitro and in vivo studies using EOs as preferred candidates for managing and treating NDS in particular AD, PD, ALS and HD in **Table 2**.

NDs	Sources of the EOs	Major constituent of the EOs	Findings	References
AD *(in Vitro)*	*Ajuga chamaecistus subsp scoparia* (Boiss.) *Rech.f*	Spathulenol (18.0%), thymol (15.1%), octen-3-ol (14.3%) and linalool oxide (11.2%).	↓ Acetylcholinesterase and butylcholestrase activities	[30]
AD *(in ivitro)*	Allium tuncelianum *(Amaryllidaceae)*	Diallyl disulfide (49.8%), diallyl trisulfide (27.9%) and allyl methyl trisulfide	↓ Lipid peroxidation activity *and* acetylcholestrase and butylcholestrase activities	[31]
AD *(in ivitro)*	*Artemisia macrocephala*	a-humulene (46.3%), (β-caryophyllene (9.3%), α-copaene (8.2%), β-myrcene (4.3%), **Z**(**E**)-α-farnesene (3.7%), and calarene (3.5%)	↓ Acetylcholestrase activities	[32]

NDs	Sources of the EOs	Major constituent of the EOs	Findings	References
AD *(in ivitro)*	*Artemisia maderaspatana*	α-humulene (46.3%), β-caryophyllene (9.3%), α-copaene (8.2%), β-myrcene	↓ Acetylcholestrase activities	[33]
		(4.3%), **Z(E)**-a-farnesene (3.7%), and calarene (3.5%)		
AD *(in ivitro)*	*Boswellia dalzielii*	3-carene (27.72%) and α-pinene (15.18%). 2,5-Dihydroxy acetophenone and β-D-xylopyranose	↓ Acetylcholestrase activities and inflammation properties	[34]
AD *(in ivitro)*	*Daucus aristidis* Coss	Contain majorly α-pinene (49–74.1%) and β-pinene (19.2–11.9%).	↓ Acetylcholestrase and butylcholestrase activities	[35]
AD *(in ivitro)*	*Panax ginseng* *Panax japonicas* *P. notoginseng* *P. quinquefolius*	Spathulenol (8.82%), bicyclogermacrene (6.23%), β-elemene (3.94%), and α-humulene (3.69%	↓ Acetylcholestrase, butylcholestrase and *β-Secretase activities.*	[36]
AD *(in vivo)*	*Lavandula angustifolia* mill.	—	↓ Acetylcholestrase activity and MDA level, ↑SOD and GPx activities. Neuroprotective effect.	[37]
AD *(in vivo)*	*Lavandula angustifolia* Mill.	—	↓ SOD and GPx activities, ↓MDA level, ↑synapse plasticity-related proteins, calcium-calmodulin-dependent protein kinase II (CaMKII), p-CaMKII, BDNF, and TrkB	[38]
AD *(in vivo)*	*Lavandula luisieri*	—	↓ BACE-1 is an aspartic protease involved in the conversion of amyloid precursor protein (APP) to Aβ.	[39]
AD *(in vivo)*	Salviae aetheroleum	—	↑ Antioxidant enzymes activity. Thus, prevent neurodegeneration.	[40]
AD *(in vivo)*	SHXW	—	↑ A/31-42 induced memory impairment and suppressed A/31-42 induced JNK, p38 and Tau phosphorylation	[41]
PD *(in ivitro)*	*Eryngium sp.*	(E)-caryophyllene (4.9–10.8%), germacrene D (0.6–35.1%), bicyclogermacrene (10.4–17.2), spathulenol (0.4–36.0%), and globulol (1.4–18.6%) are major compounds.	↓ MAO activity	[41]

NDs	Sources of the EOs	Major constituent of the EOs	Findings	References
PD *(in ivitro)*	*Cinnamomum verum Cinnamomum cassia*	—	↑ Cell viability ↓ ROS and apoptosis. Hence, neuroprotection	[42]
PD *(in ivitro)*	*Cuminum cyminum*	Cuminaldehyde	↑ Cell viability ↓*a-SN fibrillation*	[43]
PD *(in ivivo)*	*Eplingiella fruticosa*	*β-caryophyllene, bicyclogermacrene and 1,8-cineole*	↓ Catalepsy and decreased membrane lipid peroxides levels	[44]
PD *(in invivo)*	*Acorus tatarinowii Schott (Shi Chang Pu)*	—	c mRNA levels of GRP78 and CHOP, ↓ expressions of phosphorylated IER1 (p-IRE1) and XBP1	[45]
PD *(in invivo)*	*Pulicaria undulata*	Carvotanacetone	↑ Dopamine, ATP levels and glutathione striatal contents. ↓ Striatal interleukin-1β (IL-l/β), tumor necrosis factor-α (TNF-*a),* and inducible nitric oxide synthase (iNOS) and MDA	[46]
PD *(in invivo)*	SHXW		↑ Dopaminergic neurons and dopamine levels, ↑ Phosphorylation levels of cAMP-response element-binding protein	[47]

Note: ↑ Improved/increased.
↑ Decreased/inhibited.

Table 2.
The biological activities of EOs on selected NDs.

5. New delivery applications for essential oils

Till date, NDs are incurable despite many scientific efforts in drug discovery and development in NDs. Most drugs discovered successfully treat the symptoms without curing the pathogenesis or root cause of the NDs with their attendant toxicities or side effects. Hence, in NDs, most molecules are for managing the conditions and are not curative. Antioxidants are adjudged safe that serve as very potent and effective molecules to many diseases by abrogating oxidative stress implicated in most disorders.

NDs most especially affect neurons of the CNS and the brain. The nature of the architecture of the CNS accounts for the innumerable hindrances to drug delivery for effective actions. The blood-brain barrier (BBB) limits the permeability and solubility of antioxidant molecules [48]; hence the antioxidant cannot reach the target CNS. In addition, antioxidants are sometimes unstable and prone to gastrointestinal degradation [14, 49]. The BBB of the brain function as the structure that controls the movement of substances (regulatory role) in the neural microenvironment. This is the interface between the blood and neural tissue, bringing about regulation [48]. Different lines of evidence suggest that BBB breakdown contributes to the pathogenesis of NDs such as ALS, MS, PD, AD, etc. This is because BBB is highly sensitive to

OS-induced damage and distortion evident in physiological factors like neural aging, vascular disorder, molecular irregularities and anatomical pathologies [13].

The important aspect or standard for drugs or molecules is ensuring a delivery system that will circumvent the mentioned hindrances and hence promote potency, stability, specificity and safety. However, the BB is the main obstacle of CNS targeted therapies in NDs, which is addressed primarily through technology. Nanotechnology is an emerging and highly innovative field with tremendous potential in different areas, such as pharmaceuticals and medicine, thanks to distinctive physical and chemical features, such as minimal size and functionalized surface characteristics of materials [50, 51]. Nanomaterials are used in nanotechnology, which have unique physical, chemical, and biological properties due to their nanoscale dimensions (typically between 1 and 100 nm). These materials come in different sizes, shapes, compositions, surface chemical features, and hollow or solid structures, which can be adjusted to produce optical, electronic, magnetic, and biological characteristics suitable for their applications. The advantage of this nanotechnology is that it made possible the interaction of specific molecules or cellular targets by creating an excellent drug delivery system and hence targeted treatments, especially in ND conditions. This drug delivery system permits concerted multifunctional qualities such as bioactivity, targeting, imaging capabilities and gene delivery. Based on these characteristics, nanotechnology in drug delivery systems is now widely accepted.

The neuroprotective properties of EOs are attributed to their unique anti-free radical and antioxidant properties, as revealed in past research or studies. The improvement of cholinergic neuron deterioration that is eminent in most NDs conditions using EOs containing antioxidants improves cognitive function and prevents brain damage. Thus, these improved mental functions are evident in memory, attention span, planning, decision-making, judgments, speech and overall coordination [52]. EOs quickly find their way across the BBB, reaching the CNS after systemic absorption and could bring some neurological intoxications. Some molecules though lipophilic (solubility) in nature, show very poor permeability across the BBB, resulting from the active efflux mechanism in the membranes of BBB. There could be instances of improved drug accumulation that are non-target sites specific, although there is improved or increased lipid solubility in the BBB [53, 54]. Additionally, the exposure of molecules bound across the cerebral endothelial membrane to degrading enzymes [54], recognition of neuropeptide and their quick degradation by BBB itself and the reinforcement of high amount concentrations of P-glycoproteins (Pgp) that remove or prevent a range of molecules from passing across brain parenchyma [55, 56] are recognized barriers to CNS drug delivery.

Over the years, many enhanced strategies for enhanced CNS drug delivery have been developed. These strategies involve pharmaceutical manipulation, BBB disruption and other methods using nanocarriers, which would help transport molecules to target sites in the brain. Some strategies are viral vectors, polymeric nanoparticles, liposomes, dendrimers, micelles, carbon nanotubes, carbon dots and carbon nano-onions, which have been exhaustively reviewed [56]. These approaches increase therapeutic responses of natural products, including EOs, with overall effectiveness in the treatment and management by increasing bioavailability and ensuring effectiveness. For example, circumventing or overcoming BBB in NDs treatments is achievable through a nanoparticles-mediated brain drug delivery approach, as revealed in AD and HD [57–59].

Over the years, increasingly published works suggest the importance of EOs and drug delivery systems in medicine, food and pharmaceuticals. Although different applications could be adopted, this chapter focuses on the strategy of encapsulating EOs. The encapsulation is advantageous, which helps to overcome the fragility and volatility, enzymatic reactions, and preserve the biological activity that confers increased activity and decreased toxicity. The overall effect of the encapsulation EOs in drug delivery systems would be the avenue for controlled drug release, increased bioavailability and efficiency [60]. The vesicular and nanoparticles lipid–based delivery formulations vis a vis micro- and nanoemulsion, liposomes, solid lipid nanoparticles (SLN), and nanostructured lipid carriers approaches have unique characteristics.

6. Conclusion

EOs are natural products with diverse biological importance in different fields due to the inherent chemical constituents or components. The crux of this chapter is the antioxidant properties of EOs immensely contribute to the prevention, treatment and management of NDs. The multiple mechanisms of action of EOs antioxidants match or counter the multifactorial processes such as oxidative stress, neuroinflammation, excitotoxicity and others involved in NDs pathology. Although attaining the most effective therapeutics is essential, the CNS's molecule or drug delivery barrier is a significant concern in treating and managing NDs. Many approaches using modern technology supported by research evidence, no doubt, have tremendously contributed to finding near or total solutions to incurable NDS. Therefore, effort must be channeled towards discovering and developing novel EOs and drug delivery systems through innovative and holistic research and collaboration towards finding lasting solutions to NDs.

Conflict of interest

The authors declared no conflicts of interest.

Author details

Kolajo Adedamola Akinyede[1*], Habeebat Adekilekun Oyewusi[1], Oluwatosin Olubunmi Oladipo[2] and Oladimeji Samuel Tugbobo[1]

1 Department of Science Technology: Biochemistry Unit, The Federal Polytechnic, Ado Ekiti, Ekiti State, Nigeria

2 Department of Science Technology: Microbiology Unit, The Federal Polytechnic, Ado Ekiti, Ekiti State, Nigeria

*Address all correspondence to: adedamola_ak@fedpolyado.edu.ng

References

[1] Elyemni M, Louaste B, Nechad I, Elkamli T, Bouia A, Taleb M, et al. Extraction of essential oils of *Rosmarinus officinalis* L. by two different methods: Hydrodistillation and microwave assisted hydrodistillation. The Scientific World Journal. 2019;**2019**:3659432

[2] Dhifi W et al. Essential oils' chemical characterization and investigation of some biological activities: A critical review. Medicine. 2016;**3**(4):25

[3] Mohamed AA, Alotaibi BM. Essential oils of some medicinal plants and their biological activities: A mini review. Journal of Umm Al-Qura University for Applied Sciences. 2023;**9**(1):40-49

[4] Sun J et al. Chemical composition and biological activities of essential oils from six lamiaceae folk medicinal plants. Frontiers in Plant Science. 2022;**13**:919294

[5] Bakkali F et al. Biological effects of essential oils—A review. Food and Chemical Toxicology. 2008;**46**(2):446-475

[6] Machado CA et al. Steam distillation for essential oil extraction: An evaluation of technological advances based on an analysis of patent documents. Sustainability. 2022;**14**(12):7119

[7] Akinyede KA et al. Comparative study of the antioxidant constituents, activities and the GC-MS quantification and identification of fatty acids of four selected helichrysum species. Plants. 2022;**11**(8):998

[8] Rajendram R, Patel VB, Preedy VR Recommended resources for oxidative stress and dietary antioxidants in neurological diseases. In: Oxidative Stress and Dietary Antioxidants in Neurological Diseases. Cambridge, MA, USA: Academic Press; 2020. pp. 143-154

[9] Yang L, Xian D, Xiong X, Lai R, Song J, Zhong J. Proanthocyanidins against oxidative stress: From molecular mechanisms to clinical applications. BioMed Research International. 2018;**2018**:1-11. DOI: 10.1155/2018/8584136

[10] Bektašević M, Politeo O. Biological application of essential oils and essential oils components in terms of antioxidant activity and inhibition of cholinesterase enzymes. In: de Oliveira MA, de Aguiar Andrade EE, editors. Essential Oils—Advances in Extractions and Biological Applications. London, UK: IntechOpen; 2022

[11] Pisoschi AM et al. Oxidative stress mitigation by antioxidants-an overview on their chemistry and influences on health status. European Journal of Medicinal Chemistry. 2021;**209**:112891

[12] Uttara B et al. Oxidative stress and neurodegenerative diseases: A review of upstream and downstream antioxidant therapeutic options. Current Neuropharmacology. 2009;**7**(1):65-74

[13] Fisher D, Alamu OA. The relationship between the blood-brain barrier, degenerative neuropathy, and oxidative stress. In: Oxidative Stress and Dietary Antioxidants in Neurological Diseases. Amsterdam, The Netherlands: Academic Press; 2020. pp. 17-32

[14] Kumar H et al. Antioxidant functionalized nanoparticles: A combat against oxidative stress. Nanomaterials. 2020;**10**(7):1334

[15] Marino A et al. Natural antioxidant compounds as potential pharmaceutical tools against neurodegenerative diseases. ACS Omega. 2022;7(30):25974-25990

[16] Mao P et al. Mitochondria-targeted catalase reduces abnormal APP processing, amyloid β production and BACE1 in a mouse model of Alzheimer's disease: Implications for neuroprotection and lifespan extension. Human Molecular Genetics. 2012;**21**(13):2973-2990

[17] Song X, Long D. Nrf2 and ferroptosis: A new research direction for neurodegenerative diseases. Frontiers in Neuroscience. 2020;**14**:267

[18] Board PG. The omega-class glutathione transferases: Structure, function, and genetics. Drug Metabolism Reviews. 2011;**43**(2):226-235

[19] Chen JJ et al. Altered central and blood glutathione in Alzheimer's disease and mild cognitive impairment: A meta-analysis. Alzheimer's Research & Therapy. 2022;**14**(1):23

[20] Chiu H-F, Venkatakrishnan K, Wang C-K. The role of nutraceuticals as a complementary therapy against various neurodegenerative diseases: A mini-review. Journal of Traditional and Complementary Medicine. 2020;**10**(5):434-439

[21] Farkhondeh T, Yazdi HS, Samarghandian S. The protective effects of green tea catechins in the management of neurodegenerative diseases: A review. Current Drug Discovery Technologies. 2019;**16**(1):57-65

[22] Adefegha SA. Functional foods and nutraceuticals as dietary intervention in chronic diseases; novel perspectives for health promotion and disease prevention. Journal of Dietary Supplements. 2018;**15**(6):977-1009

[23] Stratakos AC, Koidis A. Methods for extracting essential oils. In: Essential Oils in Food Preservation, Flavor and Safety. Amsterdam, The Netherlands: Elsevier; 2016. pp. 31-38

[24] Soares GABe et al. Exploring pharmacological mechanisms of essential oils on the central nervous system. Plants. 2021;**11**(1):21

[25] Lourenço SC, Moldão-Martins M, Alves VD. Antioxidants of natural plant origins: From sources to food industry applications. Molecules. 2019;**24**(22):4132

[26] Benkhaira N, Koraichi S, Fikri-Benbrahim K. In vitro methods to study antioxidant and some biological activities of essential oils: A review. Biointerface Research in Applied Chemistry. 2022;**12**(3):3332

[27] Gonzalez-Burgos E, Gomez-Serranillos M. Terpene compounds in nature: A review of their potential antioxidant activity. Current Medicinal Chemistry. 2012;**19**(31):5319-5341

[28] Essa M et al. Review of natural products on Parkinson's disease pathology. The Journal of Aging Research and Clinical Practice. 2014;**3**(1):1-8

[29] Abd Rashed A, Abd Rahman AZ, Rathi DNG. Essential oils as a potential neuroprotective remedy for age-related neurodegenerative diseases: A review. Molecules. 2021;**26**(4):1107

[30] Movahhedin N et al. Ajuga chamaecistus subsp. scoparia (Boiss.) Rech. f.: A new source of phytochemicals for antidiabetic, skin-care, and neuroprotective uses. Industrial Crops and Products. 2016;**94**:89-96

[31] Karakaya S et al. Chemical and morphological characterization of Allium

tuncelianum (Amaryllidaceae) and its antioxidant and anticholinesterase potentials. Anales del Jardín Botánico de Madrid. 2019

[32] Shoaib M et al. In vitro acetylcholinesterase and butyrylcholinesterase inhibitory potentials of essential oil of Artemisia macrocephala. Bangladesh Journal of Pharmacology. 2015;**10**(1):87-91

[33] Jyotshna et al. Chemical composition and acetylcholinesterase inhibitory activity of Artemisia maderaspatana essential oil. Pharmaceutical Biology. 2015;**53**(11):1677-1683

[34] Kohoude MJ et al. Chemical composition and biological activities of extracts and essential oil of Boswellia dalzielii leaves. Pharmaceutical Biology. 2017;**55**(1):33-42

[35] Lamamra M, Laouer H, Amira S, Erdogan Orhan I, Senol F, Demirci B, et al. Chemical Composition and cholinesterase inhibitory activity of different parts of daucus aristidis coss. Essential oils from two locations in Algeria. Records of Natural Products. 2017;**11**:147-156

[36] Kawamoto H, Takeshita F, Murata K. Inhibitory effects of essential oil extracts from panax plants against β-secretase, cholinesterase, and amyloid aggregation. Natural Product Communications. 2019;**14**(10):1934578X19881549

[37] Xu P et al. Protective effect of lavender oil on scopolamine induced cognitive deficits in mice and H2O2 induced cytotoxicity in PC12 cells. Journal of Ethnopharmacology. 2016;**193**:408-415

[38] Xu P, Wang K, Lu C, Dong L, Gao L, Yan M, et al. The protective effect of lavender essential oil and its main component linalool against the cognitive deficits induced by D-galactose and aluminum trichloride in mice. Evidence-Based Complementary and Alternative Medicine. 2017;**2017**:7426538. DOI: 10.1155/2017/7426538

[39] Videira R et al. Dose-dependent inhibition of BACE-1 by the monoterpenoid 2, 3, 4, 4-tetramethyl-5-methylenecyclopent-2-enone in cellular and mouse models of Alzheimer's disease. Journal of Natural Products. 2014;**77**(6):1275-1279

[40] Cioanca O et al. In vitro–in vivo correlation of the antioxidant capacity of Salviae aetheroleum essential oil. Farmácia. 2015;**63**(1):34-39

[41] Jeon S et al. SuHeXiang wan essential oil alleviates amyloid beta induced memory impairment through inhibition of tau protein phosphorylation in mice. The American Journal of Chinese Medicine. 2011;**39**(05):917-932

[42] Ramazani E et al. Protective effects of Cinnamomum verum, Cinnamomum cassia and cinnamaldehyde against 6-OHDA-induced apoptosis in PC12 cells. Molecular Biology Reports. 2020;**47**:2437-2445

[43] Morshedi D et al. Cuminaldehyde as the major component of Cuminum cyminum, a natural aldehyde with inhibitory effect on alpha-synuclein fibrillation and cytotoxicity. Journal of Food Science. 2015;**80**(10):H2336-H2345

[44] Beserra-Filho JI et al. Eplingiella fruticosa leaf essential oil complexed with β-cyclodextrin produces a superior neuroprotective and behavioral profile in a mice model of Parkinson's disease. Food and Chemical Toxicology. 2019;**124**:17-29

[45] Ning B et al. β-Asarone inhibits IRE1/XBP1 endoplasmic reticulum

stress pathway in 6-OHDA-induced parkinsonian rats. Neurochemical Research. 2016;**41**:2097-2101

[46] Issa MY et al. Neuroprotective effects of Pulicaria undulata essential oil in rotenone model of Parkinson's disease in rats: Insights into its anti-inflammatory and anti-oxidant effects. South African Journal of Botany. 2020;**132**:289-298

[47] Quanfeng L et al. Neuroprotective effects of Suhexiang wan on the in vitro and in vivo models of Parkinson's disease. Journal of Traditional Chinese Medicine. 2019;**39**(06):800

[48] Abbott NJ et al. Structure and function of the blood–brain barrier. Neurobiology of Disease. 2010;**37**(1):13-25

[49] Hu B et al. Food macromolecule based nanodelivery systems for enhancing the bioavailability of polyphenols. Journal of Food and Drug Analysis. 2017;**25**(1):3-15

[50] Fiorito S et al. Toxicity and biocompatibility of carbon nanoparticles. Journal of Nanoscience and Nanotechnology. 2006;**6**(3):591-599

[51] Zhao H et al. Graphene-based nanomaterials for drug and/or gene delivery, bioimaging, and tissue engineering. Drug Discovery Today. 2017;**22**(9):1302-1317

[52] Lizarraga-Valderrama LR. Effects of essential oils on central nervous system: Focus on mental health. Phytotherapy Research. 2021;**35**(2):657-679

[53] Habgood M, Ek J. Delivering drugs into the brain: Barriers and possibilities. Therapeutic Delivery. 2010;**1**(4):483-488

[54] Davson H, Segal MB. Physiology of the CSF and Blood–Brain Barriers. Boca Raton: CRC Press; 1996

[55] Savolainen J et al. Effects of a P-glycoprotein inhibitor on brain and plasma concentrations of anti-human immunodeficiency virus drugs administered in combination in rats. Drug Metabolism and Disposition. 2002;**30**(5):479-482

[56] Ahlawat J et al. Nanocarriers as potential drug delivery candidates for overcoming the blood–brain barrier: Challenges and possibilities. ACS Omega. 2020;**5**(22):12583-12595

[57] Saraiva C et al. Nanoparticle-mediated brain drug delivery: Overcoming blood–brain barrier to treat neurodegenerative diseases. Journal of Controlled Release. 2016;**235**:34-47

[58] Bhatt R et al. Development, characterization and nasal delivery of rosmarinic acid-loaded solid lipid nanoparticles for the effective management of Huntington's disease. Drug Delivery. 2015;**22**(7):931-939

[59] Smith A et al. Nanolipidic particles improve the bioavailability and α-secretase inducing ability of epigallocatechin-3-gallate (EGCG) for the treatment of Alzheimer's disease. International Journal of Pharmaceutics. 2010;**389**(1-2):207-212

[60] Cimino C et al. Essential oils: Pharmaceutical applications and encapsulation strategies into lipid-based delivery systems. Pharmaceutics. 2021;**13**(3):327

Chapter 4

Methods for Evaluating the *In Vivo* Analgesic and Anti-Inflammatory Activity of Essential Oils

Mimouna Yakoubi, Nasser Belboukhari, Khaled Sekkoum, and Mohammed Bouchekara

Abstract

Essential oils (EOs) are products of the secondary metabolism of plants, and the constituents are mainly monoterpenes and sesquiterpenes of formula general $(C_5H_8)n$. The beneficial effects of the volatile compounds of essential oils have been used for a very long time by ancient civilizations to treat common pathologies. Today, so-called natural medicines are enjoying growing success with the public. Numerous studies have demonstrated that the essential oil has significant potential as antibacterial, antifungal, antioxidant, antidiabetic and painkiller agents. The use of animals is widespread in biomedical research and pain studies are no exception. Since experimentation on human subjects must be limited to moderate stimuli that do not cause injury or disease, the researchers turned to animals to find answers to their questions. Several methods have been used for the evaluation of the anti-inflammatory activity of medicinal plant extracts, and most of the tests used to evaluate pain and inflammation in animal experiments involve inducing pain in animals with different agents.

Keywords: essential oil, medicinal plants, animal experimentation, nociception, rodents, analgesic activity, anti-inflammatory activity

1. Introduction

Definitely, pain has been defined by the International Association for the Study of Pain (IASP, 1979) as an 'unpleasant sensory and emotional experience related to actual, potential, or described tissue damage by the patient' [1, 2]. Typically, it is triggered by noxious stimuli and transmitted *via* specialised neural networks to the central nervous system (CNS), where it is interpreted as such. It is also a way of protecting the body [3]. In this regard, pain and inflammation remain the most important and devastating health problems, affecting 80% of the world's population [4]. They are considered a major environmental health problem, affecting all countries. [5]. Further, untreated, persistent and prolonged pain is the most common problem, causing both physical damage and psychological distress [6]. Likewise, pain can be caused by a variety of factors, inclusive of injury, illness and psychological factors [3].

More to the point, the mechanisms of pain and inflammation, together with the modes of action of treatments, have shown to be numerous and complex. Some are still little known. Thus, treatments, classified into three levels (non-opioid, weak opioid and strong opioid), are often combined and provide relief for a large proportion of patients [7]. In this regard, analgesics or painkillers are designed to reduce or abolish painful sensations without causing loss of consciousness or suppressing other sensitivities [8]. Likewise, anti-inflammatory medicines are symptomatic drugs that do not act on the cause. They are indicated when inflammation, a normal process of defence against aggression, becomes bothersome, particularly because of pain [9]. In virtue of which, they are very widely used in a large number of diseases and more specifically in the presence of inflammation, the same as in rheumatology [3].

More and more, alternative therapies are unconventional methods of pain relief that do not involve drugs or surgery. Alternative therapies include acupuncture, massage therapy and chiropractic care. Hence, these therapies are often used in conjunction with traditional medical treatments [3]. Traditional medicine has strong cultural roots, with many plants used to treat pain. This ancestral knowledge can be regarded as a source of inspiration for finding numerous active ingredients and as a consequence enabling therapeutic innovation in the management of pain and inflammation [10].

2. Essential oils

The active molecules, involved in plant defence mechanisms, are derived from secondary metabolism. Besides, they are not directly involved in plant growth but have evolved to provide natural protection against attacks by microbes or insects [11]. Above and beyond, some of such secondary metabolites are concentrated in the oil sacs, which are essential oil-secreting pockets [12]. In consequence, exploring essential oils for molecules with biological activity seems to be an interesting avenue.

2.1 Definition

The eighth edition of the French Pharmacopoeia defines EOs as 'products of generally fairly complex composition containing volatile products contained in plants and more or less modified during preparation'. There are various processes for extracting these volatile principles. At the time, EOs were also referred to as 'essences' or 'volatile oils'. Since the ninth edition (1972), the Pharmacopoeia now only uses the term 'essential oil' [13–15]. Additionally, essential oils have shown to be complex natural mixtures of volatile secondary metabolites, isolated from plants by hydrodistillation or mechanical expression [16]; they are obtained from leaves, seeds, buds, flowers, twigs, herbs, bark, wood, roots or fruit, but alike from the gums that run off the trunks of trees. Above and beyond, hydrodistillation is still the most widely used method of producing essential oils, in particular for commercial purposes [17]. In addition, secondary metabolites are extracted from plants by steam distillation. The volume of essential oil recovered depends on the distillation yield, which varies for the same plant depending on the season [18]. Likewise, essential oils can be obtained by cold expression, as in the case of citrus fruits. New techniques have been developed to increase production yields, in respect such as extraction using liquid carbon dioxide at low temperature and high pressure [19] or extraction assisted by ultrasound or microwaves [20].

2.2 Chemical composition of essential oils

Contrary to what might be suggested by its name, a pure and natural essential oil does not contain any fats. It is made up of molecules with a carbon skeleton. As well, essential oils represent complex mixtures that may contain more than 300 different compounds [21–24]. In virtue of which, such compounds are volatile molecules, the vast majority of which belong to the terpene family. Only the most volatile terpenes, that is. those whose molecular weight is not too high, are used [23]. Terpene compounds are products of secondary metabolism. There are mainly monoterpenes (C_{10}) and sesquiterpenes (C_{15}). Several thousand compounds have been described and have shown to be classified according to their number of rings (acyclic, mono- and bicyclic compounds) together with the nature of the functions they carry (alcohols, aldehydes, ketones, ether-oxides) [25].

2.3 Physico-chemical properties of essential oils

In general, essential Oils are colourless or pale yellow in their liquid state at ordinary temperatures. However, all EOs are volatile, fragrant and flammable; their specific gravity is usually less than one [26]. Only three officinal EOs have a specific gravity greater than that of water: cinnamon, clove and sassafras. They are sparingly soluble in water, soluble in alcohols and in most organic solvents. They are alterable and are very sensitive to oxidation [27].

Essential oils share organoleptic properties (characteristics of a substance that can be perceived by the sense organs: flavour, odour, appearance and consistency of the object) such as being liquid at room temperature, volatile and water vapour permeable [28].

Modern scientific work has made it possible to better understand the essences and to precisely define their different constituents and their physico-chemical characteristics, revealing the principle of their long-known therapeutic action. The physiological role of essential oils in the herb kingdom remains unknown. However, the molecular diversity of the metabolites they contain endows them with a wide range of biological roles. Additionally, several essential oils, such as cinnamon, cayenne pepper, bay leaf and oregano oils, have antioxidant properties [29, 30].

An anti-inflammatory effect has been described for essential oils of citrus cultivars. The results suggest that *C. japonica* and *C. maxima* are promising candidates for relieving inflammatory diseases. These research findings provide the scientific basis for using essential oils from citrus cultivars to reduce inflammatory symptoms [31]. Above and beyond, the antifungal activities of numerous essential oils, including thyme, citronella, cinnamon and tea tree oils [17], have been described. The efficacy of oils extracted from yarrow, *Achillea fragrantissima* [32] and *A. milefolium* [33] against the pathogenic yeast *Candida albicans*, has also been demonstrated.

Certain essential oils have anti-tumour activity and are used in the preventive treatment of certain types of cancer. The essential oil, isolated from the seeds of *Nigella sativa* L., demonstrates cytotoxic activity *in vitro* against various tumour cell lines. *In vivo*, it limits the proliferation of liver metastases and delays the death of mice that have developed the p815 tumour [34].

2.4 Analgesic and anti-inflammatory activity of essential oils

A number of pain-relieving essential oils can help us to overcome certain ailments, in respect such as dental pain, headaches or chronic inflammation. The essential oils presented below can effectively be used to fight pain. Most of them have analgesic, analgesic and anaesthetic mechanisms of action. Lemon eucalyptus essential oil acts as an anti-inflammatory thanks to its citronellal content of over 65%, which helps to relieve pain. Clove essential oil is one of the ultimate pain-relieving essential oils. Besides, its eugenol content of over 80% gives it formidable analgesic properties. It is recommended for relieving dental pain, and peppermint essential oil has multiple properties that provide fast and effective pain relief. Firstly, it acts as a local analgesic not only thanks to its menthol content but also as a mild anaesthetic [35].

The bioactive constituents of essential oil extracted from many medicinal plants were known to provide protection against prolonged inflammation and improve the health of mortals. In this regard, the anti-inflammatory properties of these plants are of immense importance to the drug discovery process [36, 37]. Specifically, lavender essential oil has been shown to inhibit inflammation by inhibiting the nascence of TNF (tumour necrosis factor) and NF-kB (nuclear factor kappa–light chain enhancer of activated B cells) in the murine brain and human umbilical vein endothelial cells. [38]. Linalool and cinnamaldehyde present in native cinnamon leaf essential oil were found to be anti-inflammatory against endotoxin introduced into mice [39].

3. Experimental animals

Healthy adult Swiss albino mice of both sexes (20 to 35 g and 6 to 8 weeks old) are used for testing. All mice are fed commercial pellets and must have access to water ad libitum, alternately. Mice must be acclimated one week before experience in all procedures to minimise stress. All mice used in studies should be handled according to internationally accepted standard guidelines for the use of laboratory animals [39, 40].

4. Extraction of essential oils

Essential oils can be extracted by methods of steam distillation, steam and water utilised strategies to isolate these aromatic essences [41]. The best extraction method will undoubtedly be one which does not use any solvent, generates no residual waste and requires no energy to process it. Obviously, there is no other solvent more ideal than water on plants. It is available, recyclable and non-toxic. Water is already widely used in the extraction of essential oils by steam distillation [42].

5. Acute toxicity test

Characteristically, in the presence of an unknown substance, the first step in the search for pharmacological activity begins with a toxicological study and in particular, the evaluation of the lethal dose 50 (LD 50), that is. the dose, which causes the mortality of 50% of the animals. In virtue of which, increasing doses of extracts are administered to rats and mice until mortality is achieved. Although this technique is highly controversial from an ethical point of view, it nevertheless provides high-quality information:

1. Firstly, it determines the toxicity of the substance and the therapeutic margin; in other words, the ratio between the active dose and the toxic dose for the animal

species tested; this is an essential step in the use of any substance for therapeutic purposes.

2. Observation of the first symptoms of toxicity reveals the target organs; in other words, those which are preferentially affected by the toxicant; toxicity stands alike for an excellent criterion for orienting research into pharmacological activity [41].

More to the point, essential oils are not products that can be used without risk. Similarly as all natural products, 'just because it is natural does not mean it is safe for the body' [23]. The acute toxicity test for essential oil extracts can quickly be carried out using the method described by the organisation for economic cooperation and development (OECD) [43]. This test consists of administering experimental doses to the animals and monitor them for signs of poisoning, including drooling, convulsions, unusual activity, loss of consciousness, coma or even death. These observations are performed regularly for up to 48 hours [44].

6. Methods used for assessment of the analgesic and anti-inflammatory activity *in vivo*

The use of animals is widespread in biomedical research, and pain studies are no exception. As experiments on human subjects have to be limited to moderate stimuli that do not cause injury or disease, researchers have turned to animals to find answers to their questions [45].

Nociception and pain constitute a vast area of neuroscience and medical scientific research. As we know, the use of animals in scientific research has been controversial since ancient times [46], although these animal models have great merit in the advancement of biomedical sciences through their important contributions to our growing understanding pathological and biological processes [47]. Over time, various tests and models have been developed in rodents to provide fundamental and translational research tools on this subject; to study pain and try to reduce it, tests using thermal, mechanical and chemical stimuli, Hyperalgesia and pain measurements, and inflammatory or neuropathic pain models constitute one of the most important tools available to researchers in this area [48]. Preclinical therapeutic research should consequently combine pain models with nociceptive tests in order to be more relevant [49].

6.1 Methods used to study the analgesic activity *in vivo*

Most of the tests used to assess pain in animal experimentation involve inducing pain in animals using different agents [50]. Further, the tests are grouped around three basic types of pain: thermal nociception, chemical nociception and mechanical nociception. Nevertheless, in the following figure, we describe and critically analyse the most commonly used behavioural tests of nociception in animals (**Figure 1**) [49].

6.1.1 Thermal nociception

In mammalian nociceptors, noxious heat above 40°C activates thermosensitive C fibres and heat above 52°C activates A fibres [51]. In addition, tests measuring

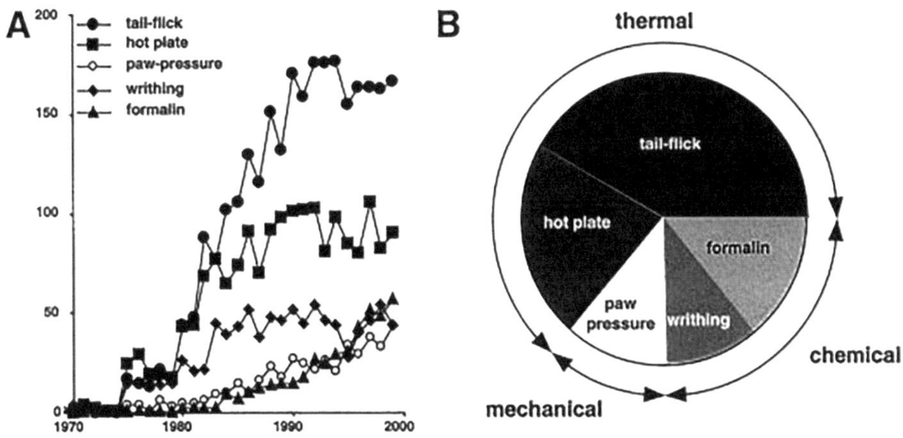

Figure 1.
A. A number of original articles published between 1970 and 1999 in which researchers used one of the five most common pain sensation tests, B. The relative proportions of these categories of articles appearing during the year 1999 (based on Medline) [49].

the nociceptive response to heat can be used experimentally in both rats and mice. Likewise, the stimulus can stop automatically when the animal responds [48, 49]. The animal is placed inside the heating plate and waited a few moments for it to react to the pain. The mouse must be recovered immediately after observing its response to the experiment to avoid any risk of burns. We can do the same experiment and repeat it several times to check the values, but this is often observed in some laboratory animals, either stress or habit of the protocol gives us different values. Repeating measurements for the same mouse may lead to different results. [49, 50]. There are also some limitations to this test, repeated measurements lead to learning phenomena, and these lead to variations in reaction latency [52].

When we touch something too hot or too cold, our senses translate this into a sensation of pain. If you put your hand in a fire, the resulting burning sensation will cause your body to move your hand away as quickly as possible. Feeling pain is actually proof that your body is working hard to keep you safe. So, losing the ability to do that means you find yourself in real trouble.

The hot plate and tail-flick tests measure an animal's ability to consciously remove a part of its body from a heat source, and they all test the ability to an animal to feel and respond to a certain degree of painful stimulation [52].

6.1.1.1 Tail withdrawal test, D'Amour and Smith test or Tail-flick *test*

It stands for a simple method that measures a spinal nociceptive reflex. The tail-flick test includes two types that are superficially similar but physically very different [53]. The first consists of immersing the animal's tail in water at a certain temperature. The second type involves applying radiant heat to a specific small area of the tail. More to the point, the surfaces stimulated can be very different. In fact of matter, it is surprising that authors generally consider these two tests to be equivalent [49]. The time taken for the animal to withdraw its tail is measured [49, 54]. Besides, to minimise the risk of tissue trauma due to exposure to heat, a time limit such as 10 seconds is set, at which point the animal is removed from the test [52].

6.1.1.2 Hot plate *test*

The hot plate method relies on measuring latency to assess skin sensitivity to pain. The response to pain usually involves licking the area to relieve the pain, shaking or immediately jumping off the hot plate [53]. Definitely, the hot plate is an open cylindrical space, the base of which is a heatable metal plate. The mouse is placed inside the plate that is preheated to a constant temperature. The animal is monitored and the response time to any type of behaviour is measured, namely paw licking and jumping. Both are considered to be integrated supraspinal responses [49, 53].

The aim was to verify the ability of the products tested to protect the animal against thermal pain. The extracts were administered either through oral or peritoneal way. The animal is placed on a heated metal plate maintained at between 52 and 55°C (**Figure 2**). The latency time for the appearance of behavioural responses is measured, with the animal licking itself, shaking its legs and/or jumping [55].

6.1.2 Chemical nociception

Chemical nociception refers to a nociceptive neuron that expresses receptors capable of detecting noxious, irritating or harmful chemicals [56, 57].

Chemociception, the detection of tissue-damaging chemicals, is important in protecting organisms from tissue damage. The ability of sensory neurons to detect potentially harmful chemicals is based on the activation of pain receptors in various animals with irritating chemical compounds. The ability of nociceptors to detect a variety of seemingly unrelated chemicals was mysterious until it was discovered that the nociceptor-specific TRPA1 ion channel (the transient receptor potential cation channel (TRP) family of receptors) could be activated by many of these chemicals [51, 56, 57].

6.1.2.1 Formalin test

Subcutaneous or intra-plantar injection of formalin into the paw of rats or mice. The time the animal spends licking and biting its paw is measured [58].

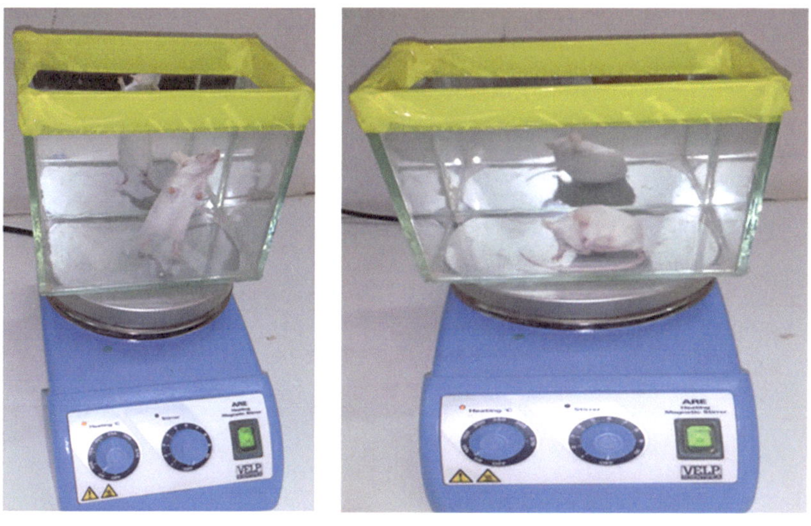

Figure 2.
The hot plate test: Latency of hind paw pain response is licking and shaking or jumping.

6.1.2.2 Abdominal contraction test, Koster test or Writhing test

The intraperitoneal administration of serosal irritants causes highly stereotyped behaviour in mice and rats [59]. We notice a decrease in the animal's motor activity, a lack of motor coordination, twisting of the dorsal abdominal muscles, abdominal contractions and the entire body (especially the hind legs), (**Figure 3**). The test goes by several names, including the abdominal torsion test, abdominal contraction response or stretch test, but it is more commonly known as the 'torsion test' [49].

In fact, the torsion test is a chemical method used to induce peripheral pain by injecting irritants, such as phenylquinone or acetic acid into mice. Further, the low frequency of contortions makes it possible to deduce the potency and analgesic activity of the active substance. Manifestations of abdominal torsion in mice were first reported by Siegmund et al. [60], as an arching of the back, extension of the hind limbs and contraction of the abdominal muscles [61]. Likewise, the number of abdominal contractions and body stretches are measured [49, 55, 59, 62, 63]. Hence, these methods have the advantage of being able to demonstrate the effects produced by weak analgesics [64].

6.1.3 Mechanical nociception

Provide responses to stimuli of excessive pressure or mechanical deformation, as well as to breaks in the skin surface (cuts, incisions). These nociceptors, which are often polymodal, have shown to be sensitive to both mechanical deformations and thermal stimuli [56].

6.1.3.1 Von Frey filaments

The von Frey test is a test of touch sensitivity using von Frey filaments, which can vary in diameter. It is considered the most important test and is almost the only mechanical test that can be used reliably not only in rats but also in mice. In rodents, they are mostly used on the plantar surface, while the animal is on a grid. The expected response is withdrawal of the paw. Rodents exhibit a withdrawal reflex when their paw is touched unexpectedly, indicating the degree of their sensitivity. Filaments of different calibres are applied [52, 64]. The von Frey test allows the

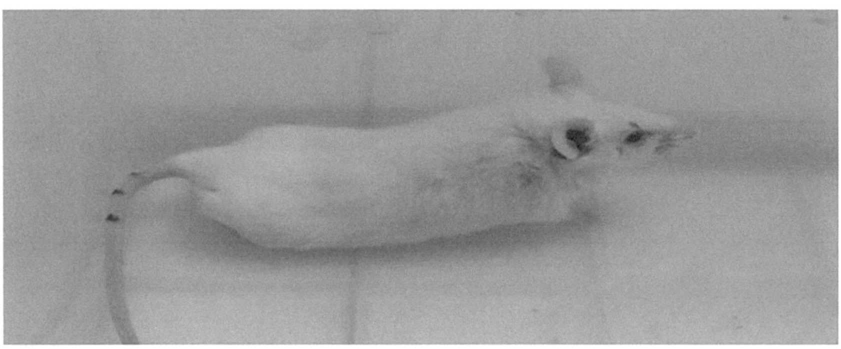

Figure 3.
Writing test: The behaviour is described as the paw stretch and contraction of the abdominal muscles.

response of the two hind legs to be differentiated, and the threshold values are stable over time, allowing repeated measurements [50, 64]. The animal's withdrawal threshold is measured in relation to the force exerted by the filament. Mechanical sensitivity test [45, 55, 64]. The von Frey bristles are nylon monofilaments, or von Frey bristles are nylon monofilaments or stiff metal bristles that exert precise levels of force when pressed against the skin. They can be used to measure mechanical stimulation.

6.1.3.2 Paw pression

The Randall and Selitto test [65] is based on determining the pain threshold induced by the application of pressure. Using a mechanical stimulator, constant or, more often, increasing pressure is applied to the animal's hind limb. The animal's behaviour is assessed: it freezes, withdraws its limb and emits cries. Electromyographic recordings of nociceptive reflexes can alike be made [49].

From all these tests, the tail withdrawal and hot plate tests remain the most commonly used. On the other hand, it should be noted that the rate of publications concerning the tail withdrawal, hot plate and torsion tests stabilised in the 1990s. In contrast, the number of articles based on the formalin test and various tests involving paw withdrawal from mechanical stimuli was noted to have increased [49].

6.1.4 Facial coding scales

A facial expression is one or more movements of the muscles or skin of the face. These movements express the emotional state of the individual to an observer. Therefore, the facial expression of pain can be used as an interesting indicator. If we can estimate these expressions quantitatively using facial coding criteria, this will enable the assessment of pain [66], and recent evidence suggests that facial expressions of pain could alike be used in rodents [67, 68]. Similarly, it is difficult to assess the internal emotional states of rodents by analysing their facial expression. Nevertheless, it has already been used, for example, to assess taste/disgust [69]. In the case of pain expression, the 'grimace scale' defined for mice [67] and rats [68] consists of noting orbital constriction. This grimace scale depends on five facial features: orbital narrowing, nose bulge, cheek bulge, ear position and moustache change. These facial action units have values of 0 (no pain), 1 (mild pain or likely pain) and 2 (severe pain or definitely present) (**Figure 4**) [68, 70]. This new approach to pain assessment in rodents could be facilitated by partial automation.

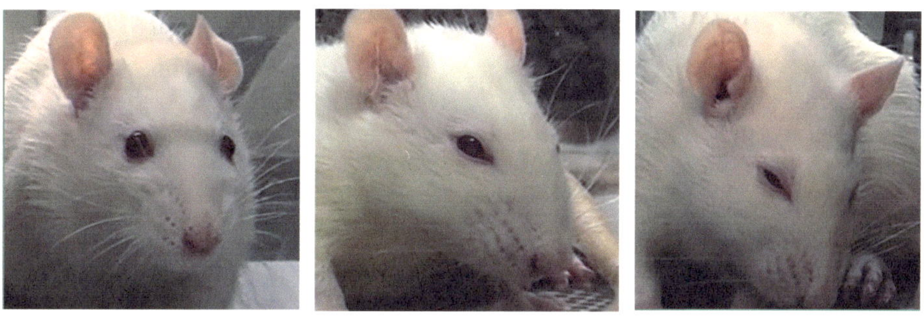

Figure 4.
Examples of visible pain expression on a rat's face, framed from the database and labelled with the rat grimace scale [48, 68] (example adapted from the Facial Pain Expression Table for Rats: Sotocinal et al., 2011).

6.2 Methods used to study the anti-inflammatory activity *in vivo*

The inflammatory response is a physiological process of defence of the body against an attack, which leads to tissue damage. The primary function of the inflammatory response is to eliminate or isolate the attacking agent (bacteria, virus, parasite, damaged tissue) from the rest of the body and to allow, as quickly as possible, tissue repair. The inflammatory response is a physiological process of defence of the body against an attack, which leads to tissue damage. The inflammatory reaction allows the elimination of the aggressors and ensures the repair of the lesions. It stops when the attacks disappear [71] and the inflammatory mediators constitute all the molecules involved in the regulation of the inflammatory process and which activate and sensitize the nociceptive system. In order to relieve inflammatory pain, the body often relies on compounds that stimulate the immune response or on inflammatory mediators themselves [72]. As a consequence, among the tests that are considered a model of short-term inflammatory pain model is the formalin test.

Several methods have been used to assess the anti-inflammatory activity of medicinal plant extracts. In virtue of which, we will illustrate some of the methods used to assess anti-inflammatory activity *in vivo*:

6.2.1 Carrageenan-induced paw oedema

Definitely, Carrageenan-induced paw oedema is certainly one of the most popular tests [72]. Therefore, it is a very sensitive and reproducible test and has been relied upon since ancient times as a model for studying new drugs effective against inflammatory pain [73]. Likewise, carrageenan-induced inflammation enables us to detect orally active acute anti-inflammatory agents. For this reason, it has great predictive value for anti-inflammatory agents that act *via* mediators of acute inflammation [74]. This inflammatory response includes three distinct phases: a first phase involving histamine and 5-hydroxytryptamine, which promote vasodilation, plasma transudation and oedema (0–1 hour), a second phase (1.5–3 hours), which uses kinins as mediators, increases vascular permeability [75] and prostaglandin biosynthesis occurs beyond the third hour (third phase) [76]. A positive effect is explained by the inhibition of the actions or synthesis of pro-inflammatory substances.

Indeed, inflammation occurs when carrageenan is injected. In rodents, intra-plantar injection of carrageenan causes hypersensitivity, which is assessed by mechanical or thermal stimulation. Pain pharmacology: The oedema caused by this photogenic agent can be translated into volume and measured, making it possible to monitor the inflammatory process (**Figure 5**) [77]. Similarly, the behaviours observed can be characterised using a scoring scale [45].

6.2.2 Croton oil-induced ear oedema

The ear oedema is induced by croton oil, using the method of Manga and colleagues. The thickness of the ear is measured using a digital calliper before treatment and a few hours after induction of inflammation [78].

6.2.3 Injection of formalin into the facial region

Injection of formalin into the rat's upper lip: The time the rat spends grooming, scratching and rubbing is assessed [79]. The injection of formalin provokes a biphasic

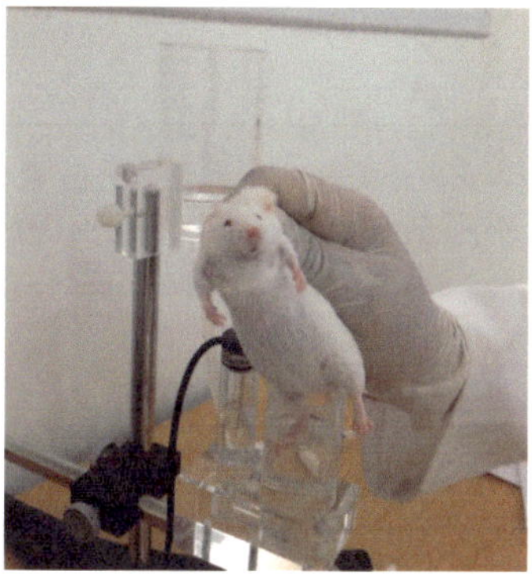

Figure 5.
Measuring the volume of the hind paw of the mice using the plethysmometer.

response similar to the one observed during intra-plantar injection. Administration of analgesics reduces nociceptive behaviour. This test can be used to assess the pain behaviour associated with trigeminal pain, as well as the effects of potentially analgesic drugs [45, 80].

6.2.4 Pulpite

The model stimulating inflammatory pain in the pulpedetar. Intradental injection of capsaicin or formalin in rats. Nociceptive behaviour is measured using a scoring scale. Model simulating inflammatory pain in the dental pulp. Nociceptive behaviour persists for approximately 2 hours. Administration of analgesic reduces the intensity thereof [45].

7. Ethical considerations animal models

If acute pain models or acute pain tests are used in which the pain does not end with the animal's response, the pain should be terminated as quickly as possible. This may mean that the animals must be humanely euthanised as soon as the test is completed (for example, convulsive test) or that analgesics must be administered. Hence, it would be preferable to use avoidance tests rather than tests in which the pain continues after the results have been obtained. Animal testing contributes to life-saving treatments for humans, and in some cases, animals must be used because it is possible that early experiments could cause catastrophic harm if conducted directly on humans. But as cruel and inhumane as animal experiments seem to be, everyone is aware that no animal leaves the laboratory alive, and during most studies, the animals are killed and ultimately dissected. This is why we, as researchers, must use animals in research and teaching, responsibly to protect animals from unnecessary pain and suffering. Researchers should also avoid exposing the animal to stress and fear, which can result from the method of conducting experiments. The number of animals involved should be as minimal as possible, and it is desirable that the duration of the experiment be as short as possible [81].

8. Conclusion

In the light of the facts enlightened above, pain is a complex and subjective phenomenon that can be caused by a variety of factors. Pain relief is an essential aspect of healthcare and is necessary to alleviate suffering and improve quality of life. Further, pain relief can be achieved through drugs, therapies and alternative therapies [3]. Traditional medicine has strong cultural roots, and many plants are used to treat pain. On the other hand, this ancestral knowledge can be the source of inspiration for finding numerous active ingredients and accordingly enabling therapeutic innovation in the management of pain and inflammation. In conclusion, many aromatic plant species have essential oils with antinociceptive activity. As a general rule, animal studies are essential for research aimed at understanding complex questions in relation to disease progression or other biological mechanisms. Since the use of human subjects in research is ethically inconceivable, the need to use animal subjects in the initial research process is essential despite the complexity of assessing pain and its manifestations, as well as measuring pain effects of potential analgesic molecules. It remains for the researcher to choose the best tests to obtain answers and identify new effective compounds with analgesic and anti-inflammatory potential.

Acknowledgements

We thank Professor Belboukhari Nacer from Tahri Mohammed University of Béchar, for the training and permission to use the laboratory instruments. His experience, his knowledge and his appetite for the field contributed greatly to the success of our research.

We appreciate the help of the staff of the Bioactive Molecules and Chiral Separation Laboratory at Béchar University.

Author details

Mimouna Yakoubi[1,2]*, Nasser Belboukhari[1], Khaled Sekkoum[1], and Mohammed Bouchekara[3]

1 Bioactive Molecules and Chiral Separation Laboratory, University of Bechar, Algeria

2 Higher Normal School of Bechar, Algeria

3 Process Engineering and Solution Chemistry Laboratory, University Mustapha Stambouli of Mascara, Algeria

*Address all correspondence to: amounyak@yahoo.fr

References

[1] Fallet V, Chaabane N, Wislez M, Cadranel J, Raffray Y. Le patient douloureux. Revue des Maladies Respiratoires Actualités. 2018;**10**:416-424. DOI: 10.1016/S1877-1203(18)30039-9

[2] Raja SN, Carr DB, Cohen M, et al. The revised International Association for the Study of Pain definition of pain: Concepts, challenges, and compromises. Pain. 2020;**161**:1976-1982. DOI: 10.1097/j.pain.0000000000001939

[3] Gilad J. Soulagement des douleurs. In: James G, editor. Mystery School; 2009. p. 208. ISBN 8887684545, 9788887684544

[4] WHO. WHO Guidelines on the Pharmacological Treatment of Persisting Pain in Children with Medical Illnesses. Geneva: World Health Organization; 2012. p. 170

[5] Calati RBC, Artero S, Ilgen M, Courtet P. The impact of physical pain on suicidal thoughts and behaviors meta-analyses. Journal of Psychiatric Research. 2015;**71**:16-32. DOI: 10.1016/j.jpsychires.2015.09.004

[6] Henschke N, Kamper SJ, Maher CG. The epidemiology and economic consequences of pain. Mayo Clinic Proceedings. 2015;**90**:139-147. DOI: 10.1016/j.mayocp.2014.09.010

[7] Coudert P, Decrozant E. Les médicaments de la douleur. Actualités Pharmaceutiques. 2021;**60**:31-35. DOI: 10.1016/j.actpha.2020.12.012

[8] Muster D. Antalgiques. Topiques. Encycl Med Chir. Ed Elsevier SAS, Paris. 2002; 6p. 22-012-G-10

[9] Weill B, Batteux F. Immunopathologie et réactions inflammatoires. Sciences médicales. Ed De Boeck Supérieur. 2003. 312 p

[10] Meunier B. L'innovation thérapeutique: évolution et tendances: Leçon inaugurale. Collège de France; 2015; 80

[11] Regnault Roger C, Philogène BJ, Vincent C. Biopesticides d'origine végétale. 2e éd ed. Lavoisier: Tec et Doc; 2008. p. 576

[12] Guinoiseau E. Molécules antibactériennes issues d'huiles essentielles [Thesis]. Universite de corse-pasquale paoli. 2010

[13] Ollier C. L'essentiel de l'aromathérapie: Le Moniteur des pharmacies. Cahier II du n°2341. 2000

[14] Pharmacopée française Xème Edition. 1994

[15] Raynaud J. Prescription et conseil en aromathérapie. Paris: Lavoisier; 2006. p. 274p

[16] Kalemba D, Kunicka A. Antibacterial and antifungal properties of essential oils. Current Medicinal Chemistry. 2003;**10**:813-829. DOI: 10.2174/0929867033457719

[17] Burt S. Essential oils: Their antibacterial properties and potential applications in foods -a review. International Journal of Food Microbiology. 2004;**94**:223-253. DOI: 10.1016/j.ijfoodmicro.2004.03.022

[18] Gonny M, Bradesi P, Casanova J. Identification of the components of the essential oil from wild Corsican *Daucus carota* L. using 13C-NMR spectroscopy.

Flavour Fragrance Journal. 2004;**19**:424-433. DOI: 10.1002/ffj.1330

[19] Santoyo S, Cavero S, Jaime L, Ibanez E, Senorans FJ, Reglero G. Chemical composition and antimicrobial activity of *Rosmarinus officinalis* L. essential oil obtained via supercritical fluid extraction. Journal of Food Protection. 2005;**68**:790-795. DOI: 10.4315/0362-028x-68.4.790

[20] Kimbaris AC, Siatis NG, Daferera DJ, Tarantilis PA, Pappas CS, Polissiou MG. Comparison of distillation and ultrasound-assisted extraction methods for the isolation of sensitive aroma compounds from garlic (*Allium sativum*). Ultrasonics Sonochemistry. 2006;**13**:54-60. DOI: 10.1016/j.ultsonch.2004.12.003

[21] Belyagoubi L. Effet de quelques essences végétales sur la croissance des moisissures de détérioration des céréales [thesis]. Université Abou Bakr Belkaïd. Tlemcen, Algeria; 2005

[22] Benayad N. Les Huiles Essentielles Extraites Des Plantes Médicinales Marocaines: Moyen Efficace De Lutte ; Contre Les Ravageurs ; Des Denrées Alimentaires Stockées [thesis]. Université Mohammed V – Agdal; 2008

[23] Piochon M. Étude des huiles essentielles d'espèces végétales de la flore laurentienne: Composition chimique, activités pharmacologiques et hémi-synthèse [thesis]. Université de Québec, Canada; 2008

[24] Rhayour K. Eessentielles sur *Esherichia coli*, *Bacillus subtilis* et sur *Mycobacterium phlei* et *Mycobacterium fortuitum* [thesis]. Université de Sidi Mohamed Ben Abdellah, Fès. Maroc; 2002

[25] Franchomme P, Jollois R, Pénoël D. L'aromathérapie exactement: Encyclopédie de l'utilisation thérapeutique des extraits aromatiques. Ed Roger Jollois ; 2001. 490p

[26] Bardeau F. Les Huiles Essentielles. Santé pratique. Ed Fernand Lanore. 2009; 315p

[27] Jacques G. Paltz SA. Le Fascinant pouvoir des huiles essentielles: précis d'hygiène vitale. Ed "Jacque Paltz". 1980; 42p

[28] Duvillard E. Les parfums: utilisations thérapeutiques et reformulation [thesis]. Universite claude bernard – L yon 1; 2013

[29] Moro Bouronzo, A. Grand guide des huiles essentielles, santé; beauté; bien-être. Ed., Hachette pratique, France. 2008; 254p

[30] Karioti A, Vrahimi-Hadjilouca T, Droushiotis D, Rancic A, Hadjipavlou-Litina D, Skaltsa H. Analysis of the essential oil of Origanum dubium growing wild in Cyprus. Investigation of its antioxidant capacity and antimicrobial activity. Planta Medica. 2006;**72**:1330-1334. DOI: 10.1055/s-2006-947255

[31] Yang J, Lee SY, Jang SK, Kim KJ, Park MJ. Anti-inflammatory effects of essential oils from the peels of citrus cultivars. Pharmaceutics. 2023;**15**:1506-1595. DOI: 10.3390/pharmaceutics15061595

[32] Barel S, Segal R, Yashphe J. The antimicrobial activity of the essential oil from *Achillea fragrantissima*. Journal of Ethnopharmacology. 1991;**33**:187-191. DOI: 10.1016/0378-8741(91)90177-F

[33] Candan F, Unlu M, Tepe B, Daferera D, Polissiou M, Sokmen A, et al. Antioxidant and antimicrobial activity of the essential oil and methanol extracts of *Achillea millefolium* subsp. millefolium Afan. (Asteraceae). Journal

of Ethnopharmacology. 2003;**87**:215-220. DOI: 10.1016/S0378-8741(03)00149-1

[34] Mbarek LA, Mouse HA, Elabbadi N, Bensalah M, Gamouh A, Aboufatima R, et al. Anti-tumor properties of blackseed (*Nigella sativa* L.) extracts. Anti-tumor effect of blackseed (*Nigella sativa* L.) extracts. Brazilian Journal of Medical and Biological Research. 2007;**40**:839-847. DOI: 10.1590/s0100-879x2006005000108

[35] Santé Douleurs. Huiles essentielles anti-douleur [Internet]. 2022. Available from: https://essentiel-des-huiles.com/sante/douleur-huiles-essentielles [Accessed: September 16, 2023]

[36] Kar S, Gupta P, Gupta J. Essential oils: Biological activity beyond aromatherapy. Natural Product Sciences. 2018;**24**:139-147. DOI: 10.20307/nps.2018.24.3.139

[37] Aoe M, Ueno-Iio T, Shibakura M, Shinohata R, Usui S, Arao Y, et al. Lavender essential oil and its main constituents inhibit the expression of tnf-α-induced cell adhesion molecules in endothelial cells. Acta Medica Okayama. 2017;**71**:493-503

[38] Lee SC, Wang SY, Li CC, Liu CT. Anti-inflammatory effect of innamaldehyde and linalool from the leaf essential oil of Cinnamomum osmophloeum Kanehira in endotoxin-induced mice. Journal of Food and Drug Analysis. 2018;**26**:211-220. DOI: 10.1016/j.jfda.2017.03.006

[39] Worlein JM. The eighth edition of the guide for the care and use of laboratory animals. implications for behavioral management. American Journal of Primatology. 2011;**73**:98

[40] Fawcett A. Guidelines for the housing of mice in scientific institutions. Animal Research Review Panel 1. Guideline 22. 2012; 143p. Animal Ethics Infolink: http://www.animalethics.org.au

[41] Fleurentin J, Cabalion P, Mazars G, Dos Santos J, Younos C. Ethnopharmacologie: sources, méthodes, objectifs. Actes de 1er colloque Européen d'Ethnopharmacologie; 23-25 mars 1990; Metz, France: ORSTOM; 1990

[42] Chemat F. In: DUNOD, editor. Eco-Extraction du Végétal: procédés innovants et solvants alternatifs. Paris; 2011. p. 336p

[43] Ejebe DE, SiminialaYI IM, Emudainohwo JOT, Ovuakporaye SI, Ojieh AE, et al. An improved technique for oral administration of solutions of test substances to experimental rats using Mediflon/Medicut intravenous cannula. African Journal of Biotechnology. 2009;**8**:960-964

[44] Paschapur MS, Patil MB, Kumar R, Patil SR. Evaluation of anti-inflammatory activity of ethanolic extract of Borassus flabellifer L. male flowers (inflorescences) in experimental animals. Journal of Medicinal, Plants Research. 2009;**3**:049-054. DOI: 10.5897/JMPR.9000611

[45] Lavigne G, Beaulieu P. Pharmacologie de la Douleur. Les Presses de l'Université de Montréal; 2005;616p

[46] Davies G, Gorman R, Greenhough B, et al. Animal research nexus: A new approach to the connections between science, health and animal welfare. Medical Humanities. 2020;**46**:499-511. DOI: 10.1136/medhum-2019-011778

[47] Lee KH, Lee DW, Kang BC. The 'R' principles in laboratory animal experiments. Laboratory Animal Research. 2020;**2020**:36-45. DOI: 10.1186/s42826-020-00078-6

[48] Barrot M. Tests and models of nociception and pain in rodents. Neuroscience. 2012;**211**:39-50. DOI: 10.1016/j.neuroscience.2011.12.041

[49] Bars D, Gozariu M, Cadden SW. Animal models of nociception. Pharmacological Review. 2001;**53**:597-652. DOI: 10.1016/j.neuroscience.2011.12.041

[50] Tracey WD. Nociception. Current Biology. 2017;**27**:R129-R133. DOI: 10.1016/j.cub.2017.01.037

[51] Elhabazi K, Ayachi S, Ilien B, Simonin F. Assessment of morphine-induced hyperalgesia and analgesic tolerance in mice using thermal and mechanical nociceptive modalities. Journal of Vision Expression. 2014;**89**:e51264. DOI: 10.3791/51264

[52] Mulder GB, Pritchett K. Rodent analgesiometry: The hot plate, tail flick and Von Frey hairs. Contemporary Topics in Laboratory Animal Science. 2004;**43**:54-55

[53] YakoubiM,BelboukhariaN,SekkoumaK, Bouchekarab M, Aboul-Eneinc HY. Analgesic and anti-inflammatory activities of *W. saharae* essential oil. Asia Pacific Journal of Molecular Biology and Biotechnology. 2021;**29**:1-10. DOI: 10.35118/apjmbb.2021.029.4.01

[54] D'amour FE, Smith DL. A method for determining loss of pain sensation. Journal of Pharmacology and Experimental Therapeutics. 1941;**72**:74-79

[55] Mogil JS, Wilson SG, Bon K, Lee SE, Chung K, Raber P, et al. Heritability of nociception I: Responses of 11 inbred mouse strains on 12 measures of nociception. Pain. 1999;**80**:67-82. DOI: 10.1016/s0304-3959(98)00197-3

[56] Macdonald AD, Woolfe G, Bergel F, Morrison AL, Rinderknecht H. Analgesic action of pethidine derivatives and related compounds. British Journal of Pharmacology and Chemotherapy. 1946;**1**:4-14

[57] Armstrong SA, Herr MJ. Physiology, Nociception. Treasure Island, FL: StatPearls Publishing; 2023

[58] Dubuisson D, Dennis SG. The formalin test: A quantitative study of the analgesic effects of morphine, meperidine, and brain stem stimulation in rats and cats. Pain. 1977;**4**:161-174. DOI: 10.1016/0304-3959(77)90130-0

[59] Koster R, Anderson M, de Beer EJ. Acetic acid for analgesic screening. Federation Proceedings. 1959;**18**:412-417

[60] Danneman PJ. Monitoring of analgesia. In: Kohn DK, Sally K, Wixson B, White WJ, John G, editors. USA: Academic Press; 1977. pp. 83-99. DOI: 10.1016/B978-012417570-9/50009-X

[61] Gawade SP. Acetic acid induced painful endogenous infliction in writhing test on mice. Journal of Pharmacology and Pharmacotherapeutics. 2012;**3**:348. DOI: 10.4103/0976-500X.103699

[62] Ness TJ. Models of visceral nociception. ILAR Journal. 1999;**40**:119-128. DOI: 10.1093/ilar.40.3.119

[63] Hogan Q. Animal pain models. Regional Anesthesia and Pain Medicine. 2002;**27**:385-401. DOI: 10.1053/rapm.2002.33630

[64] Chaplan SR, Bach FW, Pogrel JW, Chung JM, Yaksh TL. Quantitative assessment of tactile allodynia in the rat paw. Journal of Neuroscience Methods. 1994;**53**:55-63. DOI: 10.1016/0165-0270(94)90144-9

[65] Randall LO, Selitto JJ. A method for measurement of analgesic activity on inflamed tissue. Archives Internationales de Pharmacodynamie et de Thérapie. 1957;**111**:409-419

[66] Lekdioui K. Reconnaissance d'états émotionnels par analyse visuelle du visage et apprentissage machine. Synthèse d'image et réalité virtuelle [cs. GR]. Université Bourgogne Franche-Comté. 2018; 190p

[67] Langford D, Bailey A, Chanda M, et al. Coding of facial expressions of pain in the laboratory mouse. Nature Methods. 2010;7:447-449. DOI: 10.1038/nmeth.1455

[68] Sotocinal SG, Sorge RE, Zaloum A, Tuttle AH, Martin LJ, Wieskopf JS, et al. The rat grimace scale: A partially automated method for quantifying pain in the laboratory rat via facial expressions. Molecular Pain. 2011;7:55. DOI: 10.1196/1744-8069-7-55

[69] Berridge KC, Kringelbach ML. Affective neuroscience of pleasure: Reward in humans and animals. Psychopharmacology. 2008;**199**:457-480. DOI: 10.1007/s00213-008-1099-6

[70] Vidal A, Jha S, Hassle S, Price T, Busso C. Face detection and grimace scale prediction of White furred mice. Machine Learning with Applications. 2022:100312. DOI: 10.1016/j.mlwa.2022.100312

[71] Weill B. Immunopathologie et réactions inflammatoires. Ed Boeck, Sciences médicales. Série Laennec. 2003; 312p

[72] Neugebauer M, Gloeckler G, Gosling JT, Rees A, Skoug R, Goldstein BE, et al. Encounter of the Ulysses spacecraft with the ion tail of comet McNaught. The Astrophysical Journal. 2007;**667**:1262-1266. DOI: 10.1086/521019

[73] Winter CA, Risley EA, Nuss GW. Carrageenin-induced edema in hind paw of the rat as an assay for antiiflammatory drugs. Proceedings of the Society for Experimental Biology and Medicine. 1962;**111**:544-547. DOI: 10.3181/00379727-111-27849

[74] Willoughby DA. The inflammatory response. Journal of Dental Research. 1972;**51**:225-227. DOI: 10.1177/00220345720510020301

[75] Wantana R, Tassanee N, Subhadhirasakul S. Antinociceptive, antipyretic, and anti-infammatory activities of *Putranjiva roxburghii* wall leaf extract in experimental animals. Journal of Natural Medicines. 2009;**63**:290-296. DOI: 10.1007/s11418-009-0336-6

[76] Sanogo R, Maiga A, Diallo D. Activites analgesique et antiinflammatoire des extraits de maytenus senegalensis, stereospermum kuntrianum et tricrilia emetica utilisees dans le traitement traditionnel des dysmenorrhees au mali. Pharm. Méd. Trad. Afr. 2006;**XIV**:123-136

[77] Dzoyem JP, McGaw LJ, Kuete V, Bakowsky U. Chapter 9 Anti-Inflammatory and Anti-Nociceptive Activities of African Medicinal Spices and Vegetables. In: Kuete, editor. Medicinal Spices and Vegetables from Africa. Academic Press; 2017. pp. 239-270. DOI: 10.1016/B978-0-12-809286-6.00009-1

[78] Jia W, Gao WY, Cui NQ, Xiao PG. Anti-inflammatory effects of an herbal medicine (Xuan-Ju agent) on carrageenan- and adjuvant-induced paw edema in rats. Journal of Ethnopharmacology. 2003;**89**:139-141. DOI: 10.1016/s0378-8741(03)00272-1

[79] Chidiac JJ, Rifai K, Hawwa NN, et al. Nociceptive behaviour induced by dental application of irritants to rat incisors: A new model for tooth inflammatory pain. European Journal of Pain. 2002;**6**(1):55-67. DOI: 10.1053/eujp.2001.0305

[80] Manga HM, Brkic D, Marie DE, Quetin-Leclercq J. In vivo anti-inflammatory activity of Alchornea cordifolia (Schumach. & Thonn.) Müll. Arg. (Euphorbiaceae). Journal of Ethnopharmacology. 2004;**92**(2-3):209-214. DOI: 10.1016/j.jep.2004.02.019

[81] Zimmermann M. Ethical guidelines for investigations of experimental pain in conscious animals. Pain. 1983;**16**(2):109-110. DOI: 10.1016/0304-3959(83)90201-4

Chapter 5

Thymol and Its Derivatives Rich Essential Oils from *Asparagus flagellaris* and Therapeutic Applications: Review

Michael Ibok, Oluwakayode Odeja and Ejike Okpala

Abstract

Asparagus flagellaris essential oils of the leaves and roots comprised of twenty-eight (28) and nineteen (19) compounds in total, accounting 97.41 and 97.03% of the oil, respectively, were discovered using GC-MS analysis. According to reports, the EOs are a blend of terpenes, terpene derivatives, non-terpenes, thymol and its derivatives. Additionally, thymol derivatives predominated in the essential oils. When compared to the reference standards Tioconazole and Gentamicin for fungi (28 mm) and bacteria (40–35 mm), respectively, the essential oil exhibited a moderate inhibitory zone (18–10 mm) on the tested organisms. Thu, the essential oils were categorized as bacteriostatic. On the DPPH radical scavenger properties, interaction between the constituents identified in the essential oils demonstrated a distinctive a free radical scavenging activity. The chemical components of *A. flagellaris*' essential oils play a key role in both its antioxidant and antibacterial properties.

Keywords: *A. flagellaris*, essential oil, antioxidant activity, GC/MS analysis, antimicrobial activity, thymol derivatives

1. Introduction

Traditional health cures in various societies have a long history dating back to discovery of primitive drugs during their battle against natural disasters, infections, maladies and provender's. Prehistoric discovered that certain foods had specific characteristics that can help individuals recover from illnesses and live a healthier life [1]. Natural products are not simply mishaps or results of the accommodation of nature but are the characteristic articulation of the increase in the intricacy of an organism [2]. There is ongoing methodology in development of innovative drugs for human diseases from plant origin. Several plants are used to provide treatment [3]. This is especially true in underdeveloped countries where traditional medicine is still used to provide basic health care. A single plant may contain many secondary metabolites with healing activities ranging from anti-glycation, antibacterial to diuretic [4].

Plants are an extraordinary source of drugs and numerous novels organically dynamic compounds particularly used in conventional prescription for management of numerous illnesses. Plants are said to be the most unpredictable substance storage facilities of unfamiliar biodynamic compounds with undiscovered potential for use in present-day medication [5]. Plants with therapeutic and aromatic qualities have been used for millennia and are still helpful in complementary therapies today. Herbs are utilized all around the world, however they are used more frequently in developing countries. Herbs have traditionally been the main route of medication delivery [6]. Numerous fragrant plants that grow in Nigeria have been identified as potential sources of essential oils (EOs) [7]. Due to their well-established usefulness, aromatic plants that yield substantial amounts of EOs can be utilized therapeutically to treat a range of diseases. Strongly fragrant aromatic components that are obtained from aromatic plants make up EOs, a chemical composition. EOs are complex blends of bioactive ingredients with a variety of structural kinds, such as mono-, di-terpenes, sesqui-, sulfur-containing, phenolic and phenylpropanoid compounds [7]. High levels of bacterial drug resistance to existing treatments and the exorbitant costs of recent generations of antibiotics may be overcome thanks to the effective antibacterial capabilities of EOs [8]. The vast array of bioactive elements in EOs interfere with different bacteria's defense mechanisms by interacting with their cellular enzymes or cell structures, which speeds up the death of microbial cells [9]. EOs and its constituents are used in fragrances, cosmetics, medicines, aromatherapy, dentistry, sanitary products, food preservatives, agriculture and additives, and natural therapies in addition to their remarkable antibacterial characteristics. Due to the billions of dollars in yearly earnings, EOs became a much more enticing topic for both academics and industry [10].

1.1 Terpenoids, thymol and biosynthesis of thymol

The term terpenoid, isoprenoid and terpenes are frequently used interchangeably [11]. Terpenes are hydrocarbons built from isoprene units joined in head-to-tail fashion, while the terpenoids are the oxygen-containing analogues of the terpenes. They are the most diversified group of plant-derived natural products, having a broad range of biological functions. Different types of terpenoids have been isolated from plants and animals [11, 12]. Terpenoids have also been classified based on the functional groups they contain, such as hydrocabons (limonene, Myrecene, α-pinene and β-pinene), aldehydes (geranial, neral), esters (bornile acetate, methyl salicylate), alcohols (menthol, bisabolol), ketones (camphor, thujone), lactones (aesculatine, citroptene), Phenols (limol, carvacrol), ethers (1, 8-cineol) and peroxides (ascaridol).

Thymol is a naturally occurring monoterpenoid (2-isopropyl-5-methylphenol). The primary active compounds of essential identified from *Thymus vulgaris* (Lamiaceae) is an isomer of carvacrol. Thymol is a potent antiseptic that have a wide range of applications in home/personal home care products such as mouthwash, hand sanitizer, and acne medicines [13]. Treatment of gastrointestinal and respiratory problems [14–17]. *Ocimum gratissimum*, *Trachyspermum ammi*, *Carum copticum*, *Oliveria decumbens*, and *Anemopsis californica* are among the plants from which thymol has been extracted [18–22]. Thymol is suggested as a potential natural remedy in these various categories of recognized pharmaceutical uses [23]. Thymol has been used in formulations as an insecticide, fungicide, and medicinal disinfectant, among other non-pharmacological uses [24, 25]. Isopentenyl-pyrophosphate (IPP) and its isomer dimethylallyl-pyrophosphate (DMAPP), both of which are produced by the plastidic methyerythritol pathway and the cytosolic mevalonic pathway which are the five-carbon building blocks from which terpenes are biosynthesized.

The biosynthetic precursor of terpenes in the Mevalonic Acid (MVA) pathway of thymol biosynthesis (**Figure 1**) is acetyl-Coenzyme A, commonly known as activated acetic acid. It is similar to Claisen condensation and involves the coupling of two acetyl-CoAs to produce acetoacetyl-CoA, a biological equivalent of acetoacetate.

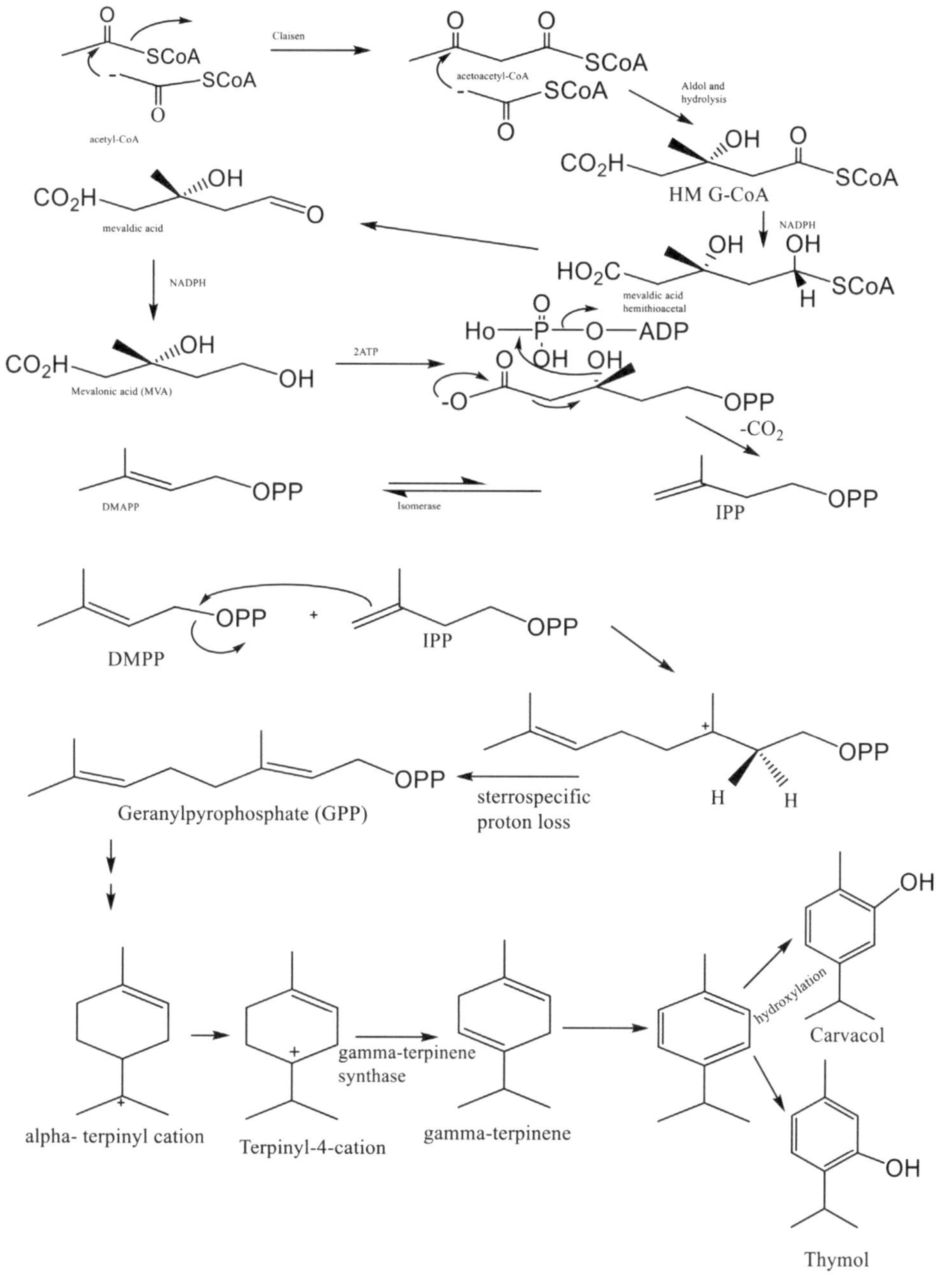

Figure 1.
Biosynthesis of thymol via the mevalonate pathway.

In a process that resembles an aldol reaction, acetoacetyl-CoA joins with another equivalent of acetyl-CoA as a carbon nucleophile to produce -hydroxy-methylglutaryl-CoA (HMG-CoA). The next step is an enzymatic reduction of HMG-CoA with nicotinamide adenine dinucleotide in the presence of water, which produces mevalonic acid [26]. Mevalonic acid monophosphate is subsequently phosphorylated by adenosine triphosphate (ATP) to produce mevalonic acid diphosphate. Isopentenylpyrophosphate (IPP), a five-carbon atom intermediate product, is then produced by decarboxylation and drying out. Isomerase, an enzyme with SH group, reacts with the IPP to produce dimethylallylpyrophosphate (DMAPP), another intermediate molecule with five carbon atoms. Geranylpyrophosphate (GPP), a precursor to monoterpenes, is produced when the electrophilic allylic CH_2 group of dimethylallylpyrophosphate (DMAPP) and the nucleophlic methylene group of isopentenylpyrophosphate (IPP) mix. Thymol and its isomer, carvacol, are produced by further reactions and rearrangements inside geranylpyrophosphate (GPP) [26].

Throughout Northern and Southern Nigeria, *A. flagellaris* (Kunth) Bak., an aromatic herb, is found [27]. Asparagaceae is its family. It has arching spiny branchlets and grows to a height of roughly 1 m as seen in **Figure 1** [28]. It is ascendant, generally erect plant. Africa as a whole uses it for a wide range of medical purposes. In both the guinea worm remedy and the ointment for hair development, the branchlets (cladodes) comprise the major component [27]. Gonorrhea, syphilis and other sexually transmitted diseases (STDs), as well as diarrhea and urinary infections, are all treated using the stem and leaves [27]. The roots are employed in Unani medicine as laxatives, tonics, aphrodisiacs, galactagogues, and treatments for renal and liver problems [29]. This herb was used to treat cholera, rheumatism and diuresis by the ancient Greeks, Romans, Indians and Chinese [30]. Indians utilized it to enhance fertility, relieve menstruation cramps, and boost nursing mothers' milk production. It works to boost kidney cellular activity, which in turn speeds up urine production. Chinese pharmacists store sparagus roots for their loved ones because they think it will make them more compassionate and loving [30]. *A. flagellaris'* stem bark and leaves were subjected to a phytochemical screening, which revealed a sizable quantity of flavonoid and a modest amount of carbohydrate, as well as trace amounts of ketones, pentose, and cardiac glycoside [28]. Mshelia *et al.* [30] reported the inhibition of six microorganisms viz. *Corynebacteria, Escherichia coli, Klebsiella, Shigella dysenteriae, Neisseria gonorrheae* and *Candida albicans* at different molarity of *A. flagellaris* ethanol extract, while the aqueous extract was susceptible to five organisms namely *Streptococcus pyogenes, Corynebacteria, Neisseria gonorrhoeae Proteus* spp. and *Treponema pallidum*.

2. Chemical constituents of *A. flagellaris*

The yield (w/w %) of the essential oils extracted from *A. flagellaris* leaves and roots using the hydrodistillation method was only 0.80 and 0.44%, respectively, based on the dry weight of the plant samples [31, 32]. The volatile oils had a distinctive aroma of thyme. The GC-MS technique was used to examine the chemicals found in the essential oils of *A. flagellaris* leaves and roots. In **Tables 1** and **2**, the detected ingredients and the total ion concentration (TIC) in % were summarized. Twenty-eight (28) and nineteen (19) compounds in total, accounting for 97.41 and 97.03% of the oil, respectively, were discovered using GC-MS analysis of leaves and root EOs. According to reports, the EOs are a blend of terpenes, terpene derivatives, non-terpenes and thymol and its derivatives.

S/N	Compounds identified	TIC (%)
1	1-(2,5-dimethoxyphenyl)-Ethanone	0.35
2	2'-Hydroxy-1'-acetonaphthone	1.77
3	2-isopropenyl-(+)-2-carene	0.24
4	8,9-dehydro-4-hydroxy thymol methylether	1.49
5	Cis-β-Farnesene	0.26
6	Durenol	3.29
7	Euparin	0.64
8	Isophosphinoline	0.76
9	n-octylesterundecanoic acid	0.98
10	Thymol methyl ether	25.8
11	Thymolhydroquinone dimethylether	12.72
12	Thymolisobutyrate	4.33
13	Thymyl tiglate	34.73
14	Thymyl-2-methylbutyrate	5.47
15	α-Bisabolol	1.36
16	α-Humulene	0.37
17	α-Santalene	0.8
18	β-Bisabolene	0.88
19	δ-Cadinene	0.79
	Total	97.03

Table 1.
Constituents of A. flagellaris *roots EO.*

Thymol and its derivatives made up the majority of the leaves essential oil (EO), which included 57.48% of it, along with its derivatives (thymol-2-methyl butyrate, 17.34%, 5-thymyl tiglate, 18.49%, thymol hydroquinone dimethyl ether, 10.52%). Additionally, the leaves essential oil (EO) has 12.99% oxygenated sesquiterpenes of therapeutic significance, including 7-epi-cis-sesquisabinene hydrate, 6-(3-isopropenyl-2-methyl-1-cyclopropen-1-yl)-6-methyl-2-heptanol and others. Germacra-4(15), 5,10(14)-trien-1-ol, cis-Farnesol, 8-Cedre-13-ol and Longiverbenone. 15.04% hydrocarbon sesquiterpenes (β-Copaene, Germacrene D, ϒ-Muurolene,δ-Cadiene, β-Bisabolene, α-uurolene, ϒ-Cadine-1,4-diene, Valence, 8,9-dihydro-cycloisolongifolene and α-Patchoulene), 6.62% oxygenated mono-terpenes (5-Propyl-1,3-benzodioxol, Bornyl acetate, (E)-Cinnamyl alcohol and 2-Allyl-cresol), 0.75% oxygenated diterpene (phytol) and 4.55% non-terpenes/non-terpenoids.

Nevertheless, the majority of the EO's roots (84.54%) were thymol derivatives (thymol hydroquinone dimethylether, 12.72%, thymol methyl ether, 25.80%, thymyl-2-methylbutyrate, thymol isobutyrate, 8, 9-dehydro-4-hydroxy thymol dimethylether, 5.4% and thymyl tiglate, 34.73%). 1.36% oxygenated sesquiterpenes (Bisabolol), 3.10% oxygenated monoterpenes (1-(2, 5-dimethoxyphenyl and Durenol)-Ethanone), 3.10% hydrogen sesquiterpenes (Santalene, Cis-Farnesene,

S/N	Compounds identified	TIC (%)
1	Thymol methyl ether	9.42
2	Bornyl acetate	0.44
3	Thymol	1.71
4	Thymol hydroquinone dimethyl ether	10.52
5	Germacrene D	3.83
6	β-Copaene	0.87
7	β-Bisabolene	0.92
8	δ-Cadinene	2.69
9	cis-Farnesol	1.51
10	ϒ-Cadine-1,4-dinene	0.37
11	α-Muurolene	0.83
12	Valencene	1.39
13	Thymyl-2-methylbutyrate	17.34
14	α-Patchoulene	2.13
15	Thymyl tiglate	18.49
16	6-(3-Isopropenyl-2-methyl-1-cyclopropen-1-yl)-6-methyl-2-heptanol	3.4
17	7-epi-cis-sesquisabinene hydrate	5.46
18	ϒ-Muurolene	2.01
19	Germacra-4(15),5,10(14)-trien-1α-ol	0.31
20	8,9-Dihydro-cycloisolongifolene	0.39
21	2-Ethylbenzimidazole	1.64
22	3-Chloro-4-fluorophenol	2.91
23	5-Propyl-1,3-benzodioxole	4.59
24	Longiverbenone	2
25	2-Allyl-p-cresol	0.37
26	(E)-Cinnamyl alcohol	1.22
27	Phytol	0.75
28	8-Cedre-13-ol	0.31
	Total	97.82

Table 2.
Constituents of A. flagellaris *leaves EO.*

Bisabolene and Humulene), and 10.99% non-terpenes/non-terpenoids. However, thymol derivatives were shown to be the main component of the EOs of *A. montana's* subsurface organs [33, 34].

3. Therapeutic application of *A. flagellaris* essential oils

Antioxidants of renown, thymol and its derivatives also have intriguing inhibitory potential against some bacteria [31, 32]. BHA and Ascorbic acid (93.09–90.40%) were competing with the essential oils of *A. flagellaris* leaves and roots, which had a substantial scavenging power from 91.04–88.06% and 90.04–50.73%, respectively (92.60–90.68%). Also noted was the concentration dependence of the antioxidant properties. The chemical compounds of the essential oil, thymol and its derivatives, which make up the majority of the mixture, may be accountable for the antioxidant property seen. However, the analysis of variance revealed no distinction in antioxidant between the essential oil of the leaves and the benchmarks that was significant ($P < 0.05$).

The essential oils from the leaves and roots of *A. flagellaris* were examined for its antibacterial properties against 10 pathogenic organisms at various concentrations (1000–62.5 μg/mL) in comparison to reference standards (gentamicin and tioconazole). The antibacterial ability of the essential oils were evaluated by measuring the inhibitory zones surrounding the well. The antibacterial activity revealed that none of the examined organisms were inhibited by the essential oils at 1000 μg/mL. *R. spp*. had the lowest inhibition at 250 μg/mL. (10 mm). The bactericidal capabilities of *A. flagellaris* leaves and roots essential oil were substantially different ($P < 0.05$) from reference standards (gentamicin and tioconazole). While only a small amount or lower concentration of the oil is needed to attack a active site in the organism, a high concentration will cause accumulation and blockage of the sensitive site and, as a result, have no effect [16]. This may explain why the oils are insensitive to the tested microorganisms at a higher concentration (1000 μg/mL).

Also, it is assumed that the protein bound at higher concentrations generating resistance to the tested microorganisms, and this was diminished as the concentration of the oil reduced. Furthermore, it has been stated that an antimicrobial's microbiologic activity is restricted to its non-protein-bound portion [31]. The interactions between the compounds present in the essential oils of the plant's leaves and roots could explain the study's findings regarding the degree of antibacterial activity. Therefore, these chemical elements may affect a number of bacterial cell target locations [35]. The essential oil's function might be regarded to as a "bacteriostatic antibiotic" since it prevents bacteria from multiplying by inhibiting bacterial protein synthesis, DNA replication, or other processes involved in cellular metabolism. This class of drugs is known as bacteriostatic antibiotics, and it includes tetracyclines, sulfonamides, spectinomycin, and others [35]. This result is in line with earlier findings on leaf extracts [6] and Marchese et al. [20] claims that all thyme essential oils are bacteriostatic.

However, those from the leaves and roots of *A. flagellaris* with noteworthy chemical compositions, such as thymol methyl ether, bornyl acetate, thymol β-bisabolene, were also found to be the most common ingredients in the root essential oil of *Chaerophyllum villosum* [36, 37]. Significant antioxidant and antibacterial activities were significantly reduced by these components. Germacrene D, farnesol, bornyl acetate and β-bisabolene have all been suggested as the possible causes of the antibacterial and antioxidant effects of *Eupatorium adenoporum* essential oil [38]. All yeasts and filamentous fungi (*F. oxysporum* f. sp. *albedinis* and *M. ramanianus*) were responsive to Thymol, Thymol Methyl Ether, α-Muurolene, Germacrene D, and α-Muurolene in *Thymus fontanesii* [39]. Thymol and thymol methyl ether were two of the key components of *Thymus vulgaris* essential oil that have antioxidant and antibacterial effects [40].

The essential oil's comparatively large content of thymol derivatives may be responsible for the majority of the reported therapeutic advantages of *A. flagellaris* in folk medicine. The literature has in-depth descriptions of strong thymol derivative antiseptic, anti-inflammatory, antibacterial, antifungal, flavoring and antispasmodic properties [38]. Other minor essential oil constituents are of phytochemical relevance because of their combined impact in suppressing the biological activities observed in this study.

4. Conclusion

The essential oils of the leaves and roots of *A. flagellaris* and their chemical makeup, antioxidant and antibacterial capabilities. The oil samples from the leaves and roots, respectively, contained 28 and 19, mostly oxygenated monoterpenes, oxygenated sesquiterpenes, hydrocarbon sesquiterpenes, oxygenated diterpenes and thymol and its derivatives. The essential oil shown significant antioxidant activity when compared to common medicines. The essential oil at a concentration between 500 and 125 μg/mL exhibited moderate activity against all the tested microorganisms, according to the inhibition zones (10–18 mm) obtained from the antimicrobial assay, in comparison to the standard drugs (gentamicin; 10 μg/mL and tioconazole; 0.7 mg/mL), which gave 26-40 mm inhibition zones. The chemical elements and biological activity of the *A. flagellaris* leaves and roots essential oil support the ethnomedicinal usage of this plant for treating syphilis, gonorrhea and other sexually transmitted disorders.

Acknowledgements

The authors appreciate the Federal University of Petroleum Resources in Effurun, Delta State, Nigeria, and University of Ibadan, Nigeria, respectively, for providing the laboratory space and equipment necessary for the extraction and analysis of the essential oils.

Conflict of interest

Oluwakayode O. Odeja, Michael Gabriel Ibok and Ejike O. Okpala declare that they have no conflict of interest.

Acronyms and abbreviations

GC/MS	gas chromatography/mass spectrometry
EOs	essential oils
TIC	total ion concentration
STDs	sexually transmitted diseases
BHA	butyl hydroxy anisole
MVA	mevalonic acid

Author details

Michael Ibok[1*], Oluwakayode Odeja[2] and Ejike Okpala[3]

1 Department of Chemistry, University of Ibadan, Ibadan, Oyo State, Nigeria

2 Department of Chemistry, Federal University of Petroleum Resources, Effurun, Delta State, Nigeria

3 Department of Chemistry, Federal University, Lokoja, Lokoja, Kogi State, Nigeria

*Address all correspondence to: michaelibok08@gmail.com

References

[1] Baliga MS, Ulloor JN, Jagetia GC, Baliga MP, Reddy R, Venkatesh P, et al. The evaluation of the acute toxicity and long-term safety of hydro-alcohlic extract of Sapthanpara (Alstonicholaris) in mice and rats. Toxicology Letter. 2004;**151**(2):317-326

[2] Jarivs BB. The role of natural products in evolution. Recent Advances in Phytochemistry. 2000;**34**:1-24

[3] Shoihawon S, Mahomadally MF. Complementary and alternative medicine use among Mauritian women. Complementary Theraipes in Clinical Practice. 2013;**2013**:36-43

[4] WHO. Prevention and Control of Schistosomiasis and Soil-Transmitted Helminthiasis. Geneva: World Health Organization; 2002

[5] Wang MY, West BJ, Jensen CJ, Nowicki D, Chen S, Palu AK, et al. Morindacitrifolia (Noni): A literature review and recent advances in Nano research. Acta Pharmacological Sinica. 2002;**23**(12):1127-1141

[6] Okwu DE, Okwu ME. Chemical composition of Spondia mombin plants. Journal of Sustainable Agriculture and Environment. 2004;**6**:140-147

[7] Vander BW. *Asparagus fagellaris* (Kunth) Baker in PROTA 2. Vegetables/legumes. Netherlands: Wageningen; 2004

[8] Burkill H. The Useful Plants of West Tropical Africa. London, United Kingdom; 1985

[9] Krtikar KR, Basu BD. Indian Medicinal Plants. Dehradun: Bishen Singh Mahendra Pal Singh; 1984

[10] British Pharmacopoeia. Drugs – Therapeutic Used and Pharmaceutical Preparations. London: British Pharmacopoeia; 1981. p. 109

[11] Withers ST, Keasling JD. Biosynthesis and engineering of isoprenoid small molecules. Applied Microbiology and Biotechnology. 2007;**73**:980-990

[12] Rohdich F, Bacher A, Eisenreich W. Isoprenoid biosynthetic pathways as anti-infective drug targets. Biochemical Society Transactions. 2005;**33**:785-791

[13] Nagoor-Meeran MF, Javed H, Al Taee H, Azimullah S, Ojha SK. Pharmacological properties and molecular mechanisms of thymol: Prospects for its therapeutic potential and pharmaceutical development. Frontiers in Pharmacology. 2017;**8**:1-34

[14] Wan L, Meng D, Wang H, Wan S, Jiang S, Huang S, et al. Preventive and therapeutic effects of thymol in a lipopolysaccharide-induced acute lung injury mice model. Inflammation. 2018;**41**:183-192

[15] Gavliakova S, Biringerova Z, Buday T, Brozmanova M, Calkovs Poliacek I, Plevkova J. Antitussive effects of nasal thymol challenges in healthy volunteers. Respiratory Physiology & Neurobiology. 2013;**187**:104-107

[16] Kissels W, Wu X, Santos RR. Short communication: Interaction of the isomers carvacrol and thymol with the antibiotics doxycycline and tilmicosin: In vitro effects against pathogenic bacteria commonly found in the respiratory tract of calves. Journal of Dairy Science. 2017;**100**:970-974

[17] Negahban M, Moharramipour S, Sefidkon F. Fumigant toxicity of

essential oil from *Artemisia sieberi* Besser against three stored-product insects. Journal of Stored Products Research. 2007;**43**:123-128

[18] Bassole I. Chemical composition and antibacterial activities of the essential oils of *Lippia chevalieri* and *Lippia multiflora* from Burkina Faso. Phytochemistry. 2003;**62**:209-212

[19] Boskabady MH, Alitaneh S, Alavinezhad A. *Carum copticum* L.: A herbal medicine with various pharmacological effects. BioMed Research International. 2014;**2014**:1-11

[20] Marchese A, Orhan IE, Daglia M, Barbieri R, Di-Lorenzo A, Nabavi SF, et al. Antibacterial and antifungal activities of thymol: A brief review of the literature. Food Chemistry. 2016;**210**:402-414. DOI: 10.1016/j.foodchem.2016.04.111

[21] Medina AL, Lucero ME, Holguin FO, Estell RE, Posakony JJ, Simon J, et al. Composition and antimicrobial activity of Anemopsis californica leaf oil. Journal of Agricultural and Food Chemistry. 2005;**53**:8694-8698

[22] Moein MR, Zomorodian K, Pakshir K, Yavari F, Motamedi M, Zarshenas MM. *Trachyspermum ammi* (L.) Sprague: Chemical composition of essential oil and antimicrobial activities of respective fractions. Journal of Evidence-Based Complementary Alternative Medicine. 2015;**20**:50-56

[23] Milovanovic S, Markovic D, Mrakovic A, Kuska R, Zizovic I, Frerich S, et al. Supercritical CO_2-assisted production of PLA and PLGA foams for controlled thymol release. Materials Science and Engineering: C. 2019;**99**:394404

[24] Benelli G, Pavela R, Canale A, Cianfaglione K, Ciaschetti G, Conti F, et al. Acute larvicidal toxicity of five essential oils (*Pinus nigra*, *Hyssopus officinalis*, *Satureja montana*, *Aloysia citrodora* and *Pelargonium graveolens*) against the filariasis vector Culex quin- quefasciatus : Synergistic and antagonistic effects. Parasitology International. 2017;**66**:166-171

[25] Tabari MA, Youssefi MR, Maggi F, Benelli G. Toxic and repellent activity of selected monoterpenoids (thymol, carvacrol and linalool) against the castor bean tick, *Ixodes ricinus* (Acari: Ixodidae). Veterinary Parasitology. 2017;**245**:86-91

[26] Dewick PM. Medicinal Natural Products, a Biosynthetic Approach. England: John Wiley & Sons Ltd.; 2002. p. 386

[27] Adams R. Identification of Essential Oil Components by Gas Chromatography/Mass Spectrometry. 5th ed. Carol Stream, IL: Texensis Publishing Gruver; 2017

[28] Olaoluwa O, Moronkola D, Taiwo O, Iganboh P. Volatile oil composition, antioxidant and antimicrobial properties of *Boerhavia erecta* L. and *Euphorbia hirta* L. TPR. 2018;**2**(3):171-178

[29] Brand-Williams W, Cuvelier ME, Berset C. Use of a free-radical method to evaluate antioxidant activity. Food Science and Technology. 1995;**28**:25-30

[30] Mshelia EH, Zaria LT, Mohammed AH, Jaji N. Phytochemical analysis and antibacterial screening of *Asparagus flagellaris* (Kunth) Bak., used in the traditional treatment of sexually transmitted diseases and urinary infections. EJESM. 2008;**1**(2):44-48

[31] Odeja OO, Ibok MG, Okpala EO. Composition and biological assays of the leaf essential oil of *Asparagus*

flagellaris (Kunth) Bak. Clinical Phytoscience. 2021;**7**:12. DOI: 10.1186/s40816-020-00245-1

[32] Odeja OO, Ibok MG, Okpala EO, Akpaeva U. Chemical composition, antimicrobial and antioxidant activities of root essential oil of Nigerian specie of *Asparagus flagellaris* (Kunth) baker. International Journal of Innovation Research & Development. 2020;**9**(5):1-6. DOI: 10.24940/ijird/2020/v9/i5/152293-375666-1-SM

[33] Andrade BFMT, Barbosa LN, Probst ID, Júnior AF. Antimicrobial activity of essential oils. Journal of Essential Oil Research. 2014;**26**(1):34-40. DOI: 10.1080/10412905.2013.860409

[34] Willuhn G. Investigation on the constituents of Arnica species VII. The composition of the essential oil from the underground organs and flower baskets of various *Arnica* species. Planta Medica. 1972;**22**:1-31

[35] Weremczuk-Jeżyna I, Wysokińska H, Kalemba D. Constituents of the essential oil from hairy roots and plant roots of *Arnica montana* L. Journal of Essential Oil Research. 2011;**23**(1):91-99

[36] Joshi RK. Root essential oil composition of *Chaerophyllum villosum* Wall. Ex DC. from Uttarakhand, India. American Journal of Essential Oils and Natural Products. 2013;**1**:34-36

[37] Dob T, Dahmane D, Benabdelkader T, Chelghoum C. Composition and antimicrobial activity of the essential oil of thymus fontanesii. Pharmaceutical Biology. 2006;**44**(8):607-612. DOI: 10.1080/13880200600897106

[38] Khadidja HB, Nadhir G, Mohamed O, Isabelle B, Mohamed Y. Chemical composition, antioxidant and antimicrobial activities of the essential oils of three Algerian *Lamiaceae* species. Current Nutrition & Food Science. 2017;**13**(2):97-109

[39] Ruberto G, Baratta M. Antioxidant activity of selected essential oil components in two lipid model systems. Food Chemistry. 2000;**69**:167-174

[40] Youmans GP, Peterson PY, Soumers HM. The Biological and Clinical Basis of Infectious Diseases. London: Longman; 1967. pp. 721-725

Section 3

Essential Oils for Animals

Chapter 6

Essential Oils in Animal Diets to Improve the Fatty Acids Composition of Meat and Milk Quality in Ruminant

Ibrahim Mohamed Khattab and Mohamed Fathy Elgandy

Abstract

Adding essential oils to the diet of ruminants is a novel strategy that improves milk and meat quality by enhancing production and fatty acid content. Including essential oils has various effects, such as modifying the biohydrogenation of unsaturated fatty acids. As a result, the fatty acid profile leaving the rumen can be affected, which in turn can affect the levels of important fatty acids in the milk and meat produced by ruminants. In the rumen, microorganisms convert unsaturated fatty acids to mostly saturated fatty acids and some unsaturated fatty acids through biohydrogenation. Added essential oils can shift the rumen microbiota, followed by changes in the fatty acid profile. The impact of essential oils on the biohydrogenation of fatty acids depends on various factors such as the type of essential oil used, its chemical composition, interactions with nutrients present in the feed, the ability of ruminal microbes to adapt to essential oils, and type of animal. Studies have indicated that various essential oils can influence ruminal fermentation and biohydrogenation of dietary fatty acids, and thus, affect the presence of polyunsaturated fatty acids in milk and meat, which is associated with positive effects on human health.

Keywords: essential oils, milk, meat, rumen microbiota, fatty acids

1. Introduction

Feed utilization is strongly affected by rumen fermentation, so nutritionists have focused on the physiology of the rumen and its microbiome. The entire rumen microbial community cooperates in the digestion and utilization of the feed, rather than the efforts of individual groups or strains to perform specific metabolic pathways such as fatty acids biohydrogenation. The shifts in the composition and abundance of ruminal microorganisms are affected not only by feed type [1] but also by the use of rumen modifiers such as plant secondary metabolites, including phenolic compounds and essential oils Essential oils (EOs) [2]. EOs are volatile, complex compounds extracted from different parts of the plants to protect plants against attacks by bacteria, fungi, or insects. Common EOs fall under two chemical groups: terpenoids and phenylpropanoids, which are synthesized through secondary metabolism in plants [3]. Chemically, extracted EOs vary due to genetic determinants of the plant, stage of growth, and environmental factors [4]. Supplementing ruminants with EOs can improve their productive performance, rumen fermentation, health, and product quality (such as milk and meat) [3, 5–7].

As a result of the prohibition of antibiotics in the animal production sector, scientists have used plant EOs and phytochemicals as available alternatives to antibiotics [8]. The EOs generally have antibacterial activity against both gram-negative and gram-positive bacteria [9, 10]. The mode of action of EOs is based on their ability to disrupt cell walls and cytoplasmic membranes, leading to lysis and leakage of intracellular compounds [11].

Many gram-positive bacteria in the rumen are involved in biohydrogenation reactions of unsaturated fatty acids (FA) [12]. Manipulating the rumen fermentation process is an important function of EOs. [13], and impeding the biohydrogenation of n-6 and n-3 fatty acids in the rumen [14], which inhibits the biohydrogenation process in the rumen and increases the passage of polyunsaturated fatty acids (PUFA) into milk and meat, and this has positive effects on human health. Extensive literature supports the use of EOs as rumen modifiers; however, more studies are urgently needed to improve the quality of milk and meat by improving their PUFA content. In this chapter, we discussed the effect of EOs on the rumen microbiome and the characteristics of dairy and meat final products.

2. Biohydrogenation in the rumen

Forages and concentrate contain abundant PUFA such as α-linolenic acid (18:3n-3), linoleic acid (18:2n-6), and oleic acid (cis-9 18:1) [15] in the form of dietary lipid. In the rumen, dietary lipids are hydrolyzed and released nonesterified FA. Then, 18:3n-3, 18:2n-6, and cis-9 18:1 are converted to saturated FA via cis-trans isomerization to trans-FA intermediates, followed by hydrogenation of the double bonds [12]. This process is called biohydrogenation. However, a small percentage of PUFA passes into milk and meat. PUFA is toxic to bacteria, so it is believed that the bacteria perform the biohydrogenation process to reduce the toxicity of PUFA [16]. Bacteria prefer saturated FA for synthesizing their cell membranes because the double bonds in unsaturated FA distort their molecular shape and disrupt the structure of the lipid bilayer [17].

In general, the bacteria responsible for biohydrogenation are classified into groups A and B [18]. Group A bacteria hydrogenate 18:2n-6 and 18:3n-3 to trans-11 18:1 and related isomers, while group B bacteria convert those 18:1 isomers to 18:0. The *Butyrivibrio* group contains the most active bacterial species involved in the biohydrogenation of C18 unsaturated FA [19, 20], including the genera *Butyrivibrio* and *Pseudobutyrivibrio*, and *Butyrivibrio proteoclasticus* [21]. Within the complex ecosystem of the rumen, some bacteria could not be exclusively classified into one of the two groups. *B. proteoclasticus*, for instance, converts 18:1 isomers not only to 18:0 but also hydrogenates 18:2n-6 to 18:0 [22].

3. Effects of EOs on rumen microbiota

The activity of EOs can alter the rumen microbiota. Early studies on the effects of dietary EOs focused mostly on the effect of EOs on feed utilization efficiency and ruminant performance. There have been only a limited number of studies conducted on the impact of EOs on rumen microbiota. These effects typically involve regulating ruminal fermentation processes thus, it affects the final products. Benchaar et al. [23] reported that the addition of EOs affected the bacteria responsible for ammonium production, the proteolytic bacteria *Provotella*, and the methane-producing archaea. EOs can reduce ruminal ammonia-N production, as they can inhibit ammonia-producing bacteria *Clostridium sticklandii*, *Peptostreptococcus anaerobius*, *lostridium sticklandii*, and *Peptostreptococcus anaerobius* and reduce the decomposition of amino acids in the rumen [24].

Depending on the type of EOs, EOs may have inhibitory effects on rumen microbiota with a decrease in rumen microbial abundance, and stimulatory effects with an increase in rumen microbial abundance [25]. EOs addition can modulate rumen fermentation by altering volatile fatty acid concentration and decreasing methane emissions by broadly altering the rumen bacterial community [26]. In beef cattle, the addition of EOs increased the relative abundance of *Parabacteroides distasonis* and *Bacteroides thetaiotaomicron* bacteria, which is associated with increased propionate concentration in the rumen [27]. Additionally, *in vitro* studies conducted by Zhou et al. [26] revealed that the addition of EOs reduced the abundance of *Succiniclasticum* bacteria, which converts the succinate to propionate, indicating a negative impact on propionate concentration. EOs can effectively reduce the presence of methanogens in the rumen, and may also have a negative impact on the relative abundance of *Ruminococcus flavefaciens*, *Ruminococcus* albus, and *Fibrobacter succinogenes* [28] and *Ruminococcus* flavefaciens in dairy goats [29]. In beef cattle, EOs have been found to increase the relative abundance of the bacterial family Succinivibrionaceae [26–30]. This increase is strongly negatively correlated with the relative abundance of Methanobacteriaceae microorganisms [31].

On the other hand, adding EOs to ruminant diets can have varying effects on the rumen microbiota. The effects are dependent on the components of the EO, which should be obtained from the same plant species, these components can also vary depending on the geographical location and the season of harvest [32]. Furthermore, the impact of the same EO on the rumen microbiota can vary depending on the dosage used. The impacts of EOs on rumen microbiota seems also more affected by ruminal pH. The pH level can affect the dissociation of EO molecules and their final form. Some studies suggest that the undissociated hydrophobic form of the active EO molecules is more effective in antimicrobial activity as it dissolves better in the bacterial membrane's lipid bilayer, this effect has been observed in low rumen pH conditions [33, 34].

4. Effect of EOs on the microbiota operating the biohydrogenation of fatty acids

Altering rumen populations with EOs may also have an effect on some rumen microorganisms responsible for the biohydrogenation of PUFA, which causes a modification of the rumen FA profile that passes into the milk and meat of ruminants [35]. The use of EO supplements suitable for ruminants must be well planned in terms of the type of EOs and dosage. A proper choice has a positive effect on feed efficiency, digestion, rumen fermentation, and meat and milk production, while an inappropriate choice can lead to negative effects, including reducing feeding efficiency and increasing methane production and resistance of rumen bacteria to the effects of EOs. The antibacterial properties of EOs may cause bacteria to become resistant to EOs. However, very few studies have researched bacterial resistance to EOs. EOs are a mixture of chemical compounds that have antimicrobial activities [8], making it hard for bacteria to develop resistance mechanisms [36]. However, it has been found that Staphylococcus aureus has developed resistance to EOs [37, 38].

5. Effects of EOs on milk fatty acids profile

Milk fat content and FA composition are primarily determined by lipid metabolism in the rumen and the mammary gland [39]. EO supplementation can reduce the levels of C8 and C12 fatty acids. This occurs because EOs can produce potent inhibitors that hinder the synthesis of de novo FA in the mammary gland [40, 41].

The rate and extent of FA biohydrogenation could also impact variations among animal species [42]. In small ruminants, supplementation of EOs is more effective in the biohydrogenation process. For instance, Juniper EO addition enhanced the presence of n-3 fatty acid in goat milk, while anise, clove, and juniper had minimal impact on milk components [43]. Cinnamon oil elevated the concentration of unsaturated FA, n3 linolenic acid, and conjugated linoleic acid [44]. The inclusion of garlic or juniper EO in the diet of dairy cows resulted in an increased proportion of conjugated linoleic acid in milk [45]. Clove and thyme EO were found to be more effective in decreasing the biohydrogenation process and increasing the concentration of C18:3 n-3, C18:4 n-3, and n-3 FA in milk fat and decreased the concentrations of C18:3 n-6, C20:4 n-6, and n-6 FA in the milk of lactating goats [6].

Dairy cattle fed a blend of EOs (capsaicin, carvacrol, cinnamaldehyde, and eugenol) secreted more CLA, suggesting that EO influenced the ruminal biohydrogenation of PUFA [46]. Supplementation with 750 mg of a mixture of EOs had no change in milk FA profile [47]. However, supplementing 2 g/d of the same mixture increased the concentration of CLA in milk fat [48].

6. Effects of EOs on meat fatty acid profile

Meat typically has a high content of saturated fatty acids (SFA) and a low content of PUFA/SFA [49]. Therefore, enhancing the PUFA levels in animal diets, either by incorporating a source rich in n-6 or n-3 PUFA, or by supplementing with EOs to inhibit biohydrogenation, generally leads to an improvement in the PUFA/SFA ratio. It has been found that the meat obtained from animals that are fed on pasture has a lower n-6/n-3 ratio compared to the meat from animals that are fed on grains because of the higher levels of α-linolenic acid found in pasture [50].

There are limited studies on the impact of supplementation of EOs to beef on its FA composition. Supplementing rosemary oil had a limited effect on the FA composition in various tissues and did not result in a significant alteration in the overall profile [51–53]. However, according to Nieto et al. [54] rosemary oil has a significant impact on the level of C18:0 content while causing a decrease in C16:0 fatty acid. The addition of anise, clove, and thyme EO enhanced the ratio of PUFA to SFA. It also increased the levels of n-3 fatty acids while decreasing the levels of n-6 fatty acids [5].

7. Conclusion

Changes in the FA profile of meat and milk resulting from EO addition have been shown to be variable and unpredictable in terms of n-3 fatty acids. Furthermore, to enhance our understanding in this area, it seems important to precisely characterize the effects of various environmental elements on the major rumen microbial species involved in the biohydrogenation process and to study how these effects are influenced by diet and ruminal environment. Further studies are needed on the factors that influence the biohydrogenation bacteria in the rumen so that we can propose new rational dietary modifications that will ultimately lead to healthier ruminant products for human consumption.

Conflict of interest

The author declares no conflict of interest.

Author details

Ibrahim Mohamed Khattab[1]* and Mohamed Fathy Elgandy[2]

1 Department of Animal and Fish Production, Faculty of Desert and Environmental Agriculture, Matrouh University, Matrouh, Egypt

2 Department of Agriculture, Faculty of Environmental Sciences, King Abdulaziz University, Jeddah, Saudi Arabia

*Address all correspondence to: imkhattab@mau.edu.eg

References

[1] Grilli DJ, Fliegerová K, Kopečný J, Lama SP, Egea V, Sohaefer N, et al. Analysis of the rumen bacterial diversity of goats during shift from forage to concentrate diet. Anaerobe. 2016;**42**:17-26

[2] Vasta V, Luciano G. The effects of dietary consumption of plants secondary compounds on small ruminants' products quality. Small Ruminant Research. 2011;**101**:150-159

[3] Casamiglia S, Busquet M, Cardoza P, Castillejos L, Ferrett A. Essential oils as modifiers of rumen microbial fermentation: A review. Journal of Dairy Science. 2017;**90**:2580-2595

[4] Cosentino SC, Tuberoso CI, Pisano B, Satta ML, Mascia V, Arzedi E, et al. In-vitro antimicrobial activity and chemical composition of Sardinian thymus essential oils. Letters in Applied Microbiology. 1999;**29**:130-135

[5] El-Essawy AM, Abdou AR, Khattab IM, Abdel-Wahed AM. Effect of addition of anise, clove and thyme essential oils on barki lambs performance, digestibility, rumen fermentation, carcass characteristics and intramuscular fatty acids. Egyptian Journal of Nutrition and Feeds. 2019;**22**:465-477

[6] El-Essawy AM, Anele UY, Abdel-Wahed AM, Abdou AR, Khattab IM. Effects of anise, clove and thyme essential oils supplementation on rumen fermentation, blood metabolites, milk yield and milk composition in lactating goats. Animal Feed Science and Technology. 2021;**271**:11ya4760

[7] Nehme R, Andrés S, Pereira RB, Ben Jemaa M, Bouhallab S, Ceciliani F, et al. Essential oils in livestock: From health to food quality. Antioxidants. 2021;**10**:330

[8] Benchaar C, Greathead H. Essential oils and opportunities to mitigate enteric methane emissions from ruminants. Animal Feed Science and Technology. 2011;**166**:338-355

[9] Hong EJ, Na KJ, Choi IG, Choi KC, Jeung EB. Antibacterial and antifungal effects of essential oils from coniferous trees. Biological and Pharmaceutical Bulletin. 2004;**27**:863-866

[10] Rota C, Carraminana JJ, Burillo J, Herrera A. In vitro antimicrobial activity of essential oils from aromatic plants against selected foodborne pathogens. Journal of Food Protection. 2004;**67**:1252-1256

[11] Nazzaro F, Fratianni F, De Martino L, Coppola R, De Feo V. Effect of essential oils on pathogenic bacteria. Pharmaceuticals. 2013;**6**:1451-1474

[12] Harfoot CG, Hazlewood GP. Lipid metabolism in the rumen. In: Hobson PN, Stewart CS, editors. The Rumen Microbial Ecosystem. London, UK: Chapman & Hall; 1997. pp. 382-426

[13] Patra AK. Effects of essential oils on rumen fermentation, microbial ecology and ruminant production. Asian Journal of Animal and Veterinary Advances. 2011;**6**:416-428

[14] Giannenas I, Skoufos J, Giannakopoulos C, Wiemann M, Gortzi O, Lalas S, et al. Effects of essential oils on milk production, milk composition, and rumen microbiota in Chios dairy ewes. Journal of Dairy Science. 2011;**94**(11):5569-5577

[15] Ferlay A, Bernard L, Meynadier A, Malpuech-Brugère C. Production of trans and conjugated fatty acids in dairy

ruminants and their putative effects on human health: A review. Biochimie. 2017;**141**:107-120

[16] Maia MR, Chaudhary LC, Bestwick CS, Richardson AJ, McKain N, Larson TR, et al. Toxicity of unsaturated fatty acids to the biohydrogenating ruminal bacterium, Butyrivibrio fibrisolvens. BMC Microbiology. 2010;**10**:1

[17] Keweloh H, Heipieper HJ. Trans unsaturated fatty acids in bacteria. Lipids. 1996;**31**:129-137

[18] Kemp P, Lander DJ. Hydrogenation in vitro of -linolenic acid to stearic acid by mixed cultures of pure strains of rumen bacteria. The Journal of General Microbiology. 1984;**130**:527-533

[19] Paillard D, McKain N, Rincon MT, Shingfield KJ, Givens DI, Wallace RJ. Quantification of ruminal clostridium proteoclasticum by real-time PCR using a molecular beacon approach. Journal of Applied Microbiology. 2007;**103**:1251-1261

[20] Durmic Z, McSweeney CS, Kemp GW, Hutton P, Wallace RJ, Vercoe PE. Australian plants with potential to inhibit bacteria and processes involved in ruminal biohydrogenation of fatty acids. Animal Feed Science and Technology. 2008;**145**:271-284

[21] Moon CD, Pacheco DM, Kelly WJ, Leahy SC, Li D, Kopecny J, et al. Reclassification of Clostridium proteoclasticum as Butyrivibrio proteoclasticus comb. nov., a butyrate-producing ruminal bacterium. International Journal of Systematic and Evolutionary Microbiology. 2008;**58**(9):2041-2045

[22] Huws SA, Lee MR, Muetzel SM, Scott MB, Wallace RJ, Scollan ND. Forage type and fish oil cause shifts in rumen bacterial diversity. FEMS Microbiology Ecology. 2010;**73**:396-702

[23] Benchaar C, McAllister TA, Chouinard PY. Digestion, ruminal fermentation, ciliate protozoal populations, and milk production from dairy cows fed cinnamaldehyde, quebracho condensed tannin, or Yucca schidigera saponin extracts. Journal of Dairy Science. 2008;**91**:4765-4777

[24] McIntosh FM, Williams P, Losa R, Wallace RJ, Beever DA, Newbold CJ. Effects of essential oils on ruminal microorganisms and their protein metabolism. Applied and Environmental Microbiology. 2003;**69**:5011-5014

[25] Caroprese M, Ciliberti MG, Marino R, Santillo A, Sevi A, Albenzio M. Essential oil supplementation in small ruminants: A review on their possible role in rumen fermentation, microbiota, and animal production. Dairy. 2023;**4**:497-508

[26] Zhou R, Wu J, Lang X, Liu L, Casper DP, Wang C, et al. Effects of oregano essential oil on in vitro ruminal fermentation, methane production, and ruminal microbial community. Journal of Dairy Science. 2020;**103**:2303-2314

[27] Zhang F, Li B, Ban Z, Liang H, Li L, Zhao W, et al. Evaluation of origanum oil, hydrolysable tannins and tea saponin in mitigating ruminant methane: In vitro and in vivo methods. Journal of Animal Physiology and Animal Nutrition. 2021;**105**:630-638

[28] Cobellis G, Trabalza-Marinucci M, Yu Z. Critical evaluation of essential oils as rumen modifiers in ruminant nutrition: A review. Science of the Total Environment. 2016;**545**:556-568

[29] Mavrommatis A, Skliros D, Flemetakis E, Tsiplakou E. Changes in

the rumen bacteriome structure and enzymatic activities of goats in response to dietary supplementation with Schizochytrium spp. Microorganisms. 2021;**9**:1528

[30] Wang Z, Li X, Zhang L, Wu J, Zhao S, Jiao T. Effect of oregano oil and cobalt lactate on sheep in vitro digestibility, fermentation characteristics and rumen microbial community. Animals. 2022;**12**:118

[31] McCabe MS, Cormican P, Keogh K, O'Connor A, O'Hara E, Palladino RA, et al. Illumina MiSeq phylogenetic amplicon sequencing shows a large reduction of an uncharacterised Succinivibrionaceae and an increase of the Methanobrevibacter gottschalkii clade in feed restricted cattle. PLoS One. 2015;**10**:e0133234

[32] Vokou D, Kokkini S, Bessiere JM. Geographic variation of Greek oregano (Origanum vulgare ssp. hirtum) essential oils. Biochemical Systematics and Ecology. 1993;**21**:287-295

[33] Cardozo PW, Calsamiglia S, Ferret A, Kamel C. Screening for the effects of natural plant extracts at different pH on in vitro rumen microbial fermentation of a high-concentrate diet for beef cattle. Journal of Animal Science. 2005;**83**:2572-2579

[34] Spanghero M, Zanfi C, Fabbro E, Scicutella N, Camellini C. Effects of a blend of essential oils on some end products of in vitro rumen fermentation. Animal Feed Science and Technology. 2008;**145**:364-374

[35] Parvar R, Ghoorchi T, Kashfi H, Parvar K. Effect of Ferulago angulata (Chavil) essential oil supplementation on lamb growth performance and meat quality characteristics. Small Ruminant Research. 2018;**167**:48-54

[36] Briskin DP. Medicinal plants and phytomedicines. Linking plant biochemistry and physiology to human health. Plant Physiology. 2000;**124**:507-514

[37] Brul S, Coote P. Preservative agents in foods: Mode of action and microbial resistance mechanisms. International Journal of Food Microbiology. 1999;**50**:1-7

[38] Nelson RR. Selection of resistance to the essential oil of Melaleuca alternifolia in Staphylococcus aureus. Journal of Antimicrobial Chemotherapy. 2000;**45**:549-550

[39] Hanuš O, Samková E, Křížová L, Hasoňová L, Kala R. Role of fatty acids in milk fat and the influence of selected factors on their variability—A review. Molecules. 2018;**23**:1636

[40] Baumgard LH, Corl BA, Dwyer DA, Sæbø A, Bauman DE. Identification of the conjugated linoleic acid isomer that inhibits milk fat synthesis. American Journal of Physiology-Regulatory, Integrative and Comparative Physiology. 2000;**278**:R179-R184

[41] Bauman DE, Griinari JM. Nutritional regulation of milk fat synthesis. Annual Review of Nutrition. 2003;**23**:203-227

[42] Zhu Z, Hang S, Zhu H, Zhong S, Mao S, Zhu W. Effects of garlic oil on milk fatty acid profile and lipogenesis-related gene expression in mammary gland of dairy goats. Journal of the Science of Food and Agriculture. 2013;**93**:560-567

[43] Morsy TA, Kholif SM, Matloup OH, Abdo MM, El-Shafie MH. Impact of anise, clove and juniper oils as feed additives on the productive performance of lactating goats. International Journal of Dairy Science. 2012;**7**:20-28

[44] Kholif SM, Morsy TA, Abdo MM, Matloup OH, El-Ella AA. Effect of supplementing lactating goats rations with garlic, cinnamon or ginger oils on milk yield, milk composition and milk fatty acids profile. Journal of Life Sciences. 2012;**4**:27-34

[45] Yang WZ, He ML. Effects of feeding garlic and juniper berry essential oils on milk fatty acid composition of dairy cows. Nutrition and Metabolic Insights. 2016;**9**:NMI-S33395

[46] Silva RB, Pereira MN, Araujo RC, Silva WD, Pereira RA. A blend of essential oils improved feed efficiency and affected ruminal and systemic variables of dairy cows. Translational Animal Science. 2020;**4**:182-193

[47] Benchaar C, Petit HV, Berthiaume R, Ouellet DR, Chiquette J, Chouinard PY. Effects of essential oils on digestion, ruminal fermentation, rumen microbial populations, milk production, and milk composition in dairy cows fed alfalfa silage or corn silage. Journal of Dairy science. 2007;**90**:886-897

[48] Benchaar C, Petit HV, Berthiaume R, Whyte TD, Chouinard PY. Effects of addition of essential oils and monensin premix on digestion, ruminal fermentation, milk production, and milk composition in dairy cows. Journal of Dairy Science. 2006;**89**:4352-4364

[49] Jeronimo E, Alves SP, Dentinho MT, Martins SV, Prates JA, Vasta V, et al. Effect of grape seed extract, Cistus ladanifer L., and vegetable oil supplementation on fatty acid composition of abomasal digesta and intramuscular fat of lambs. Journal of Agricultural and Food Chemistry. 2010;**58**:10710-10721

[50] Enser M, Hallett KG, Hewett B, Fursey GA, Wood JD, Harrington G. Fatty acid content and composition of UK beef and lamb muscle in relation to production system and implications for human nutrition. Meat Science. 1998;**49**:329-341

[51] Morán L, Giráldez FJ, Panseri S, Aldai N, Jordán MJ, Chiesa LM, et al. Effect of dietary carnosic acid on the fatty acid profile and flavour stability of meat from fattening lambs. Food Chemistry. 2013;**138**:2407-2414

[52] Smeti S, Hajji H, Mekki I, Mahouachi M, Atti N. Effects of dose and administration form of rosemary essential oils on meat quality and fatty acid profile of lamb. Small Ruminant Research. 2018;**158**:62-68

[53] Güney M, Karaca S, Erdogan S, Kor A, Kale C, Onalan S, et al. Effects of dietary supplementation with rosemary oil on methanogenic bacteria density, blood and rumen parameters and meat quality of fattening lambs. Italian Journal of Animal Science. 2021;**20**:794-805

[54] Nieto G, Díaz P, Bañón S, Garrido MD. Dietary administration of ewe diets with a distillate from rosemary leaves (*Rosmarinus officinalis* L.): Influence on lamb meat quality. Meat Science. 2010;**84**:23-29

Chapter 7

Oregano Essential Oils as a Nutraceutical Additive in Poultry Diets

Liliana Betancourt López

Abstract

Oregano essential oils (OEOs) are volatile compounds obtained from the leaves of the oregano plant (*Origanum vulgare*) through a process of steam distillation. Its major bioactive compounds include carvacrol and thymol. The OEOs from *Origanum* spp. have been considered to be the best ones because they have a higher content of carvacrol; however, in tropical America and Africa ecosystems, chemotypes with a high thymol content are found as *Lippia origanoides*. Carvacrol and thymol are responsible for the strong antimicrobial, antioxidant, and anti-inflammatory effects of OEO. They have been studied for their potential effects on the function of intestinal barrier, which plays a crucial role in maintaining gut health. These compounds may help reduce inflammation in the gut, protect and preserve intestinal integrity, control the growth of harmful bacteria in the gut, and improve the poultry's productive performance. The OEO has potential as a natural alternative to antibiotics for improving intestinal health and barrier function in poultry. However, more research is needed to fully understand the mechanisms underlying these effects and determine the optimal dosage and duration of OEO supplementation.

Keywords: *Lipia origanoides*, *Origanum*, thymol, carvacrol, growth promoter, nutraceuticals

1. Introduction

The *Origanum* genus covers a wide range of more than 60 plant species, most of which belong to the *Lamiaceae* and *Verbenaceae* families; among these species, *Origanum vulgare* L., *Lippia graveolens* Kunth, and *Lippia origanoides* Kunt are considered species of economic importance with a wide range of applications [1, 2]. Oregano essential oils (OEOs) are volatile compounds obtained from oregano leaves by steam distillation; they are composed of more than 50 phytochemically different molecules, such as terpenes, phenols, alcohols, organic acids, aldehydes, and ketones. Those metabolites are responsible for the medicinal properties of OEO, including antibiotic and anti-inflammatory properties that can help strengthen the immune system of birds [3]. The OEO may improve feed intake, the growth rate of broilers, and the overall body condition and quality of meat carcasses [4, 5]. Oregano is also known to fight off bacteria that frequently make poultry sick, such as *Escherichia coli*, *Salmonella*, and *Clostridium* [2, 6]. The OEO is now used commercially by producers in place of antibiotic feed. Due to its antibiotic properties, oregano has been found to improve intestinal functions and alter the gut microbiota in broiler chickens and laying hens. Another application of OEO has been the improvement of intestinal antioxidant capacity, immunity, and intestinal microbiota in birds [7, 8]. Recently, the effects of oregano essential oil (OEO) on intestinal health and barrier function in poultry have been investigated. OEO has been shown to improve the barrier function of intestinal epithelium and consequently strengthen immune defense against pathogens in laying hens [8]. This mechanism appears to be based on regulating intestinal bacteria and inflammation. OEO has been shown to improve intestinal antioxidant capacity, immunity, and gut microbiota in yellow-feathered chickens and, therefore, improve growth performance, antioxidant status, and intestinal health in broilers. In pigs, the OEO has also been found to improve intestinal morphology and expression of tight junction (TJ) proteins associated with modulation of selected intestinal bacteria and immune status [9]. Furthermore, feeding OEO to pullets during rearing can help improve flock uniformity, reproductive fitness, and feed efficiency [8]. The OEO is known for its potent antimicrobial properties, which are primarily attributed to its major bioactive compounds, including carvacrol and thymol. These compounds possess significant biological activities and have been extensively studied for their antimicrobial, antioxidant, anti-inflammatory, and anticancer properties [10].

The previous evidence suggests that OEO has potential as a natural alternative to antibiotics for improving intestinal health, barrier function, and productive performance in the poultry industry. This chapter presents some experiences that we have had, as well as those of other authors, about the uses and applications of the different OEO chemo types for poultry production.

2. Oregano essential oils composition

When working with locally grown OEO, there is a wide range of secondary metabolite compositions that determine its wide variation in the biological activity. In our studies, 54 compounds were identified in the volatile fraction of *Origanum* and *Lippia* genus essential oils. Carvacrol and thymol and their precursors p-cymene and ɤ-terpinene were the majority of phenolic monoterpenoid compounds found. *Origanum vulgare* L. ssp. Hirtum, grown in Greece, showed the highest value of carvacrol (90.3%), in contrast to the major volatile compounds found in *O. vulgare* L. Grown under greenhouse conditions was thymol (21.51%). Carvacrol showed the lowest value (4.3%). In contrast, *Lippia origanoides* Kunth, an endemic species of the Patia region [2], presented the highest value of thymol (78.7%). *L. origanoides* is very important for Colombia because it is an endemic species of arid tropical lands [11]. *L. origanoides* is a native species of oregano adapted to arid conditions, dry soil in the Patia region of Colombia's southwest [12]. These conditions allow that *Lippia origanoides* Kunth is a unique species because of its high thymol content (close to 80%), low variation in metabolites, and low percentage of precursor compounds. *Origanum majorana* was another species cultivated under a greenhouse. It was represented by the bicyclic monoterpenoids sabinene and cis- and trans-sabinene hydrates (17.1%), with a high level of precursors ɤ-terpinene and terpinen 4-ol (20.0%), very low thymol (10.0%), and carvacrol (3.7%), with the presence of sesquiterpenoid compounds such as germacrene and bicyclogermacrene, acyclic monoterpenoid b-myrcene, and sesquiterpenoid trans-β-caryophyllene [2]. In contrast, in other tropical areas, it has been found that carvacrol is the major component, together with ɤ-terpinene, thymol, and p-cymene, in oils from *L. origanoides* leaf hat [13]. In contrast, in our study, carvacrol was present only as a trace constituent (0.9%), but thymol contents ranging from 1.4 to 74.4% are reported [14] due to new chemotypes found. It can be concluded that the crop, the weather, the species, and the seasonal conditions can affect secondary metabolite production [15].

Despite the great variability, European and Greek oreganos have been the most evaluated and processed to be distributed on a commercial scale throughout the world because they have the highest content of carvacrol [13]. However, to tropical ecosystems such as America and Africa ecosystems, where *Lippia* genus native species with high thymol content is found, it might be important. Additionally, carvacrol has been shown to have more negative effects than thymol [16], raising the question of whether a high carvacrol content in the essential oil is a good indicator of its quality.

3. Antimicrobial activity of OEO

The OEO's antimicrobial activity has been attributed to its compounds such as carvacrol and thymol, which are the main constituents responsible for the strong antimicrobial effects of oregano essential oil [2, 6]. They exhibit broad-spectrum activity against various bacteria, fungi, and even some viruses. These compounds disrupt the cell membranes and inhibit the growth and proliferation of microorganisms [10, 17]. The OEO damage disrupts in-membrane integrity, which affects pH homeostasis and equilibrium of inorganic ions [18]. Carvacrol has been shown to alter the electrolyte balance because of decrease in the intracellular potassium, increase in the extracellular potassium, and consequently the bacterial membrane potential [19]. But the effect of OEO goes beyond the bacterial membrane because a transfer of monoterpene to inside of bacteria and the interaction with their components is also suggested [20]. Another mechanism of antibacterial action has been found for the OEO, the inhibition of quorum sensing (QS) in *Escherichia coli* and *Pseudomonas aeruginosa*. The QS is a mechanism that modulates the expression of genes that allow to recognize the bacterial population density through the accumulation of specific signaling molecules to ensure the survival and pathogenesis [21]. The OEO has proven to have a wide spectrum of antibacterial activity and is one of the most inhibitory species. However, there are few references that compare the antimicrobial activity of *Origanum* gender species versus *Lippia origanoides* as well as the antibacterial activity against pathogenic and beneficial bacterial strains to animal health and production.

Our studies compared the minimum bactericidal concentration (MBC) of carvacrol, *O. vulgare* spp., and *L. origanoides*. Carvacrol presented the lowest value of MBC against *E. coli* (0.006 mg/mL) and *Salmonella typhimurium* (0.098 mg/mL), followed by *L. origanoides* against *E. coli*, and *O. vulgare* L. ssp. Hirtum and *O. vulgare* L. against *S. typhimurium*. The highest bactericidal activity against *S. enteritidis* was presented with carvacrol, *O. vulgare* L. ssp. Hirtum (high carvacrol), and *L. origanoides* (high thymol). In contrast, the lowest bactericidal activity was presented against beneficial bacteria with a higher value of MBC when they were compared with pathogenic bacteria. Except to *Bifidobacterium breve*, while carvacrol presented the highest bactericidal activity (1.56 mg/mL), L. *origanoides* had the lowest activity (50 mg/mL). The *O. majorana* essential oil showed the lowest bactericidal activity against pathogenic and beneficial bacteria. The higher bactericidal activity against pathogenic bacteria was associated with the content of phenolic components carvacrol or thymol with respect to species such as *O. majorana* with a high content of sabinyl compounds. This study found that the OEO had a selective bactericidal effect against pathogenic and beneficial bacteria. The beneficial bacteria *Lactobacillus* and *Bifidobacterium* were less sensitive to OEO than the pathogens *Salmonella* and *E. coli* to both carvacrol and OEO high carvacrol and high thymol. These results give to OEO a potential use to maintain the intestinal eubiosis [12]. Other studies also concluded that the essential oil of *Lippia origanoides* H.B.K. presented antimicrobial action against *Staphylococcus aureus*, *Escherichia coli*, and *Candida albicans* strains [22]. The OEO from *Lippia origanoides* high thymol grown in Brazil also demonstrated antimicrobial activities against the foodborne bacteria *Bacillus cereus*, *B. subtilis*, and *Salmonella typhimurium*, but not against *Pseudomonas aeruginosa* [23]. It can be concluded that *L. Origanoides* presents a potential for the development of new phytotherapeutics.

4. Benefits as an additive in poultry feed

The food-animal industry has used antibiotics in the feed for decades. The use of antibiotics in subtherapeutic doses in animal feed has been questioned due to its possible relationship with resistance in bacteria that has an impact on human and animal health and the antibiotic resistance is one of the top health challenges of this century [24]. In recent years, there has also been a growing interest in antibiotic-free broiler meat production, therefore, it is necessary to generate nonantibiotic alternatives for their use in feed and animal nutrition. Bioactive compounds derived from medicinal plants have been proposed as natural sources of phytochemicals with antibacterial, anti-inflammatory, and antiviral properties, among others, which are ideal for use as feed additives in food animal production [25, 26]. Within these phytochemicals, oregano essential oils have been evaluated and used as additives in poultry feed. Various studies have verified the positive effects of OEOs on feed intake, metabolism, digestive secretions, and growth, among others [4, 27, 28], as well as nonculturable pathogens [29]. Both studies and commercial products with OEO have been based on essential oils from Greek oregano species (*Origanum* spp.) with high carvacrol content. It has been proposed that the effectiveness of these OEOs may be the result of their antimicrobial activity as presented back. Our studies on *Lippia origanoides* Kunth, also commonly named oregano, have shown that AEO rich in thymol can be a viable natural additive to produce chicken meat with a possible anticoccidial effect. We found that the inclusion level of AEO depends on the broiler's health status. This species, native to tropical America, was analyzed by gas chromatography, and 22 compounds were identified in the volatile fraction of the essential oil. We established that thymol was the major secondary metabolite (78.7%) in *L. origanoides* Kunth with a very low level of precursors such as p-cymene (6.6%), ɤ-terpinene (2.7%), and carvacrol (0.9%); however, this type of oregano has been little studied. Our studies included the effect of OEO from *L. Origanoides* Kunth on cecal microbial diversity and composition in broilers challenged with attenuated coccidia oocyte vaccine (2×) and found that supplementation with 100 ppm of AEO reduced the negative impact of the challenge with attenuated oocysts of coccidia. A significant interaction was identified, because 65 ppm was OEO level that maximized the body weight in non-coccidian-challenged chicken groups, but 147 ppm was required for the coccidian-challenged group. The OEO supplementation to coccidia-challenged broilers improved the body weight (9.3%) and the feed conversion ratio (FCR) (-6%). It was shown that AEO rich in thymol can be a viable natural additive to produce chicken meat without antibiotic [30].

Average daily gain (ADG) and feed conversion ratio (FCR) were better in broilers fed with OEO than those fed with a control diet during days 1–21 and days 22–42 of age [3]. Although there is no agreement on the inclusion levels, 150 or 300 mg/kg of OEO has proven to be effective in increasing the average daily feed intake and average daily gain [31]. Other studies raised higher levels of 150–750 ppm of OEO in broiler diets and concluded that during the grower period, the OEO-supplemented groups showed a better conversion of feed, with linear and quadratic performance. A linear increase in body weight, body weight gain, the protein efficiency ratio, the relative growth rate, and better conversion was observed at 750 ppm of OEO dietary level compared to the control groups. With the broken-line regression model, it was found that 580 ppm of OEO optimizes the final body weight and the feed conversion [32].

We compared OEO from four chemotypes against antibiotics in broiler diets and, contrarily, *O. majorana* with slow carvacrol, thymol, and antibiotics presented higher values of energy and fat ileal digestibility and a higher body weight. Interestingly, a negative correlation between the body weight and carvacrol intake (r: −0.55) and a positive correlation with thymol intake (r: 0.46, p 0.05) have been reported. These results found a possible negative effect of high carvacrol intake and different responses of broilers as a function of OEO composition [33]. The impact of OEO on digestive efficiency had been more consistent, with a greater amylase enzyme activity, a quadratic elevation in chymotrypsin, and higher chymotrypsin and lipase activity, which would be expressed in a better efficiency in the use of food in broilers, greater average daily gain (ADG) and a lower fructose-to-glucose (F/G) ratio [8], a linear increase in villus height to crypt depth ratio, a quadratic decrease in feed conversion ratio with a lower feed conversion ratio (200 ppm) than the control during weeks 1–12 of the trial, and higher eggshell thickness at the end of weeks 4, 8, and 12 [8]. With a level of 275 ppm of OEO in the diet of laying hens, it improved the body weight, lowered the triglycerides, and raised the high-density lipoprotein levels [34]. The OEOs, or secondary metabolites, have been very widely used in the poultry industry in mixtures with other types of essential oils with beneficial effects on body weight and feed conversion ratio values when evaluated in coccidia-challenged broiler chickens [35, 36]. Coccidia is an important parasite because it generates a high economic impact and welfare and becomes a challenge for the poultry industry. On the other hand, the use of drugs as feed additives is being withdrawn, thus, these OEO mixtures as well as *L. origanoides* essential oils are a potential option as a natural alternative to antibiotics for improving the poultry productive performance. However, more research is needed to fully understand the mechanisms underlying these effects and determine the optimal dosage and duration of OEO supplementation.

5. Effects on gut microbiota

The gut microbiota is considered one of the key elements contributing to the efficiency, welfare, and health of birds. The evaluation of gut microbiota changes in response to the use of feed additives is a key indicator to evaluate the efficiency of natural alternatives to the use of feed antibiotics in poultry production. The latest generation molecular techniques have contributed to characterizing the microbial communities in a complex environment such as the digestive tract [33, 37, 38] and have also allowed them to transcend to the animal's metabolome. In our studies, a positive correlation was found between body weight and OEO from *L. origanoides* supplementation with the Firmicutes:Bacteriodetes ratio and Firmicutes levels in cecal content. Quite fewer Bacteroidetes bacteria than those of Firmicutes have been found in fat broilers [39], and a similar relationship was found in obese humans experiencing insulin resistance [40]. The addition of 200 ppm of dietary OEO increased the abundances of Burkholderiales, *Actinobacteria*, Bifidobacteriales, Enterococcaceae, and Bacillaceae, whereas it decreased *Shigella* abundance in the ileum [8].

We also explored the effects of OEO from three chemotypes through the use of denaturing gradient gel electrophoresis (DGGE) technology on broiler chickens at high altitude, *Origanum vulgare* L. (OL), *O. vulgare* L. ssp. Hirtum (OH), and *O. majorana* (OM) from a greenhouse of Sabana de Bogotá and *O. vulgare* L. ssp. Hirtum (OG) from Greece. Bacterial community DNA revealed two main clusters: OEO-treated chicks and nontreated control chicks. These results indicate that all treatments with OEO had some effect on the gut microbial communities' changes. A reduction of about 50% in mortality caused by Ascites with *O. Majorana* essential oil and an average of 68% of all additive-supplemented groups compared to controls indicate a possible association of pulmonary hypertension syndrome with the use of OEO and gut microbiota [41].

It is also reported that OEO supplementation increases the abundances of *Ruminococcus*, *Bifidobacterium*, and *Enterococcus* as well as its Spearman's correlation positively correlated with the messenger RNA (mRNA) expression of mucins. Moreover, the relative abundance of *Enterococcus* was positively correlated with amylase activity [42]. The OEO supplementation enriched the genera *Propionibacterium*, *Brevinema*, and *Corynebacterium*, whereas the genus *Vibrio* was enriched in the control with no OEO supplementation in *Cyprinus carpio* [3].

6. The OEO and tight junctions as a barrier between intestinal cells

The intestinal barrier is formed by a layer of epithelial cells with intercellular junction complexes that form a regulated, selectively permeable barrier between luminal contents and the underlying tissue compartments. These include tight junction (TJ) proteins, such as claudins (CLDN), occludins (OCL), and *zonula occludens* (ZO), which form the continuous intercellular barrier between epithelial cells required to regulate the selective permeability across the intestinal epithelium to a barrier function [43, 44]. Although there is limited research about oregano's effects on tight junctions in birds, some studies have investigated the potential benefits of its active compounds on intestinal barrier function. Dietary OEO (150–300 mg/kg) increased the content of secretory immunoglobulin A and the relative expression of Claudin 1, Mucin 2, and Avian β-defensin 1 in yellow-feathered chicken's ileum [31]. The OEO has been shown to improve intestinal morphology and expression of tight junction proteins, which are associated with the regulation of selected intestinal bacteria and immune status in pigs supplemented with 25 mg/kg of OEO for 4 weeks. These results show that OEO promotes intestinal barrier integrity, probably through modulation of intestinal bacteria and immune status in pigs [9]. The OEO supplementation in pig decreased serum endotoxin levels, increased the villus height and expression of the TJ zonula occludens-1 (ZO-1), signaling pathways and expression of inflammatory cytokines such as mitogen-activated protein kinase (MAPK), protein kinase B (Akt), and nuclear factor kappa B (NF-kB) in the jejunum. A lower population of *Escherichia coli* in the jejunum, ileum, and colon, and inactivation of inflammation were observed too. The supplementation of OEO in production hens' late phase resulted in a quadratic reduction in the mRNA expression of interleukin 1 (IL-1), tumor necrosis factor alpha (TNF-α), interferon gamma (IFN-γ), and toll-like receptor-4 (TLR-4), but a linear and quadratic increase of ZO-1 expression in the ileum. It was observed that the addition of 200 mg/kg of OEO maximized the results [8]. The effect of OEO on sheep was studied too and it showed that this effect reduces the serum interleukin 2 (IL-2) and TNF-β levels as well as mRNA levels of NF-kB, p65, toll-like receptor-4 (TLR-4), and interleukin 6 (IL-6) in the jejune mucosa [42]. All these studies show that the integrity of the intestinal barrier was improved by OEO treatment.

As mentioned earlier, carvacrol and thymol, two major components of oregano essential oil, have been studied for their antimicrobial and anti-inflammatory properties. These compounds have been shown to inhibit the growth of certain pathogenic bacteria and reduce inflammation in the gut. By modulating the gut microbiota and reducing inflammation, oregano compounds might indirectly support the integrity of tight junctions. Additionally, oxidative stress has been implicated in the disruption of tight junction proteins, potentially compromising the intestinal barrier. The OEO antioxidant effects could counteract the oxidative stress, thus reducing the oxidative stress, and may help protect tight junction integrity [45]. However, it's important to note that the research on oregano's effects on tight junctions and the intestinal barrier is limited, and more studies are needed to fully understand its mechanisms and potential benefits.

7. The OEO antioxidant properties

The potent antioxidant properties of OEO have also been attributed to these phenolic compounds: carvacrol, thymol, and rosmarinic acid. They scavenge free radicals, neutralize oxidative stress, and protect cells from damage caused by reactive oxygen species (ROS). The OEO supplementation in broiler chickens improved the antioxidant indices in serum, glutathione peroxidase (GSH-Px), superoxide dismutase (SOD), and glutathione reductase (GR) on day 21 and the activity of total antioxidant capacity (T-AOC) of birds on day 42 [3]. Similar results were observed in jejunum, decreasing serum oxidative stress parameters, increasing the activity of T-AOC, and decreasing of the level of malondialdehyde (MDA) in serum and jejunum. The oregano leaf-flower oils were shown by the 2,2-diphenylpicrylhydrazyl (DPPH) free radical scavenging assay that they had the strongest antioxidant activities, while the stem oils had the weakest, and although the essential oils of each part showed strong antioxidant activities, their antioxidant capacities were much lower than those of the synthetic antioxidant butylated hydroxytoluene (BHT) [7]. In contrast, oregano essential oils from *Lippia origanoides* Kunt and butylated hydroxytoluene (BHT) had similar antioxidant capacities, which can be attributed to the synergistic action of OEO phenolic compounds [46]. Even more, it is reported that the radical scavenging activity of the essential oil from *Lippia origanoides* was higher than that of BHT but similar to that of α-tocopherol [45]. And *Lippia berlandieri* Schauer demonstrated the preservation of ground beef quality to be similar to that of BHT [47].

8. Anti-inflammatory properties

The anti-inflammatory effect of OEO has been verified; as mentioned before, the OEOs inhibit the production of pro-inflammatory cytokines and enzymes, thereby reducing inflammation in various tissues and organs, which has been demonstrated in pigs [9]. The OEO downregulates the expression of TNF-α, interleukin 1β (IL-1β), interleukin 6 (IL-6), IFN-γ, and monocyte chemoattractant protein-1 (MCP-1) and inhibited the activation of c-Jun N-terminal kinase (JNK), extracellular signal-regulated kinase 1/2 (ERK1/2), and protein kinase B (Akt) in pig jejunum. The greater inactivation of inflammation was reflected by the mitogen-activated protein kinase (MAPK), protein kinase B (Akt), and nuclear factor kappa B (NF-kB) signaling pathways and expression of inflammatory cytokines in the jejunum; downregulated the relative expression of tumor necrosis factor α (TNF-α) and transforming growth factor β (TGF-β) as well as a significant increase in nitric oxide, viability, and differentiation into neutrophil-like cells.

9. Other applications

Carvacrol and thymol have demonstrated cytotoxic effects against various cancer cell lines, including breast, colon, and prostate cancer cells. This activity has been explained by the management of inflammatory conditions of OEO. The OEO metabolites induce apoptosis (programmed cell death), inhibit tumor growth, and show promise as adjuvants in cancer treatment. Furthermore, it has been suggested that OEO oil metabolites, particularly carvacrol, may have antidiabetic effects. They help regulate blood sugar levels, improve insulin sensitivity, and inhibit certain enzymes involved in carbohydrate metabolism [10]. However, further research is needed to establish their clinical significance in diabetes management.

Other applications in ethnopharmacological studies have been found to suggest *L. origanoides* for disorders of the genitourinary system used specially by quilombola women from Oriximiná. These applications have also been found to demonstrate an antispasmodic action, analgesic and antimicrobial uses of other species of *Lippia* genus rich in thymol and carvacrol [48]. In addition, oregano has been found to modulate innate immunity parameters in *Lumbricus terrestris* [21]. It has also been shown to improve immune response, activity of digestive enzymes, and intestinal microbiota of the koi carp, *Cyprinus carpio* [49].

10. Conclusions

The European oreganos are *Origanum* spp. have been considered the best ones because they have a higher content of carvacrol; however, in tropical America and Africa ecosystems, chemotypes with a high thymol content, such as *Lippia Organoides* spp., it has been shown that they have functional effects on poultry health and production.

Oregano essential oils have many benefits for poultry, including strengthening the immune system, inhibiting pathogenic bacterial growth, improving intestinal integrity, improving egg quality, and promoting growth performance. It is important to note that the effectiveness of oregano may vary depending on the dosage and form of administration.

Our studies found that *Lippia origanoides* Kunth is a promising species due to its high thymol content (close to 80%), low variation in active compounds, and low percentage of precursors. It also demonstrated strong antimicrobial activity against pathogenic bacteria and the lowest antimicrobial activity against beneficial bacteria. Additionally, the OEO supplementation of coccidia-challenged broilers improved their productive performance, showing that *L. Origanoides* EO high thymol can be a viable natural additive to produce chicken meat.

Consequently, OEO has potential as a natural alternative to antibiotics for improving intestinal health and barrier function in poultry. However, more research is needed to fully understand the mechanisms underlying these effects and determine the optimal dosage and duration of OEO supplementation.

In human health, it is necessary to investigate other applications of the OEO for anticancer, anti-inflammatory, and antidiabetic properties.

Acknowledgements

The authors thank Agrosavia, Cundinamarca, Colombia, where the bioassays were performed, the Ministry of Agriculture and Rural Development by the sponsorship of the US Department of Agriculture, Universidad Nacional de Colombia, and Universidad de La Salle.

Conflict of interest

The author declares no conflict of interest.

Author details

Liliana Betancourt López
Universidad Nacional de Colombia, Bogotá, D.C., Colombia

*Address all correspondence to: llbetancourtl@unal.edu.co

References

[1] Padulosi S, Leanman D, Quek P. Challenges and opportunities in enhancing the conservation and use of medicinal and aromatic plants. Journal of Herbs Spices & Medicinal Plants. 2002;**9**:243-267. DOI: 10.1300/j044v09n04_01

[2] Betancourt L, Phandanauvong V, Patiño R, Ariza-Nieto C, Afanador-Téllez G. Composition and bactericidal activity against beneficial and pathogenic bacteria of oregano essential oils from four chemotypes of Origanum and Lippia genus. Revista de la Facultad de Medicina Veterinaria y de Zootecnia. 2012;**59**:21-31. Available from: http://www.redalyc.org/articulo.oa?

[3] Zhang LY, Peng QY, Liu YR, Ma QG, Zhang JY, Guo YP, et al. Effects of oregano essential oil as an antibiotic growth promoter alternative on growth performance, antioxidant status, and intestinal health of broilers. Poultry Science. 2021;**100**:101-163. DOI: 10.1016/j.psj.2021.101163

[4] Peng QY, Li JD, Li Z, Duan ZY, Wu YP. Effects of dietary supplementation with oregano essential oil on growth performance, carcass traits and jejunal morphology in broiler chickens. Animal Feed Science and Technology. 2016;**214**:148-153. DOI: 10.1016/j.anifeedsci.2016.02.010

[5] Hernandez-Coronado AC, Silva-Vazquez R, Rangel-Nava ZE, Hernandez-MartinezCA,Kawas-GarzaJR, Hume ME, et al. Mexican oregano essential oils given in drinking water on performance, carcass traits, and meat quality of broilers. Poultry Science. 2019;**98**:3050-3058. DOI: 10.3382/ps/pez094

[6] Leal ALAB, Braga AL, de Araújo BB, Rodrigues AMX, de Carvalho Teixeira FAF. Antimicrobial action of essential oil of *Lippia origanoides* H.B.K. Journal of Clinical Microbiology and Biochemical Technology. 2019;**5**:007-012. DOI: 10.17352/jcmbt.000032

[7] Fei H, Guang-qiang MA, Yang M, Yan L, Xiong W, Shu J-c, et al. Chemical composition and antioxidant activities of essential oils from different parts of the oregano. Journal of Zhejiang University-Science B (Biomedicine & Biotechnology). 2017;**18**(1):79-84. DOI: 10.1631/jzus.B1600377

[8] Feng J, Mingyuan L, Wang J, Zhang H, Qiu K, Qi G, et al. Dietary oregano essential oil supplementation improves intestinal functions and alters gut microbiota in late phase laying hens. Journal of Animal Science and Biotechnology. 2021;**12**:2-15. DOI: 10.1186/s40104-021-00600-3

[9] Yi Z, Xiang Q, Wang J, Peng J, Wei H. Oregano essential oil improves intestinal morphology and expression of tight junction proteins associated with modulation of selected intestinal bacteria and immune status in a pig model. BioMed Research International. 2016;**2016**:5436738. DOI: 10.1155/2016/5436738

[10] Nagoor Meeran MF, Javed H, Al Taee H, Azimullah S, Ojha SK. Pharmacological properties and molecular mechanisms of thymol: Prospects for its therapeutic potential and pharmaceutical development. Frontiers in Pharmacology. 2017;**8**:380. DOI: 10.3389/fphar.2017.00380

[11] Vega-Vela NE, Delgado-Avila WA, Chacon-Sanchez MI. Genetic structure and essential oil diversity of the

aromatic shrub *Lippia origanoides* Kunth. (Verbenaceae) in two populations from northern Colombia. Agronomía Colombiana. 2013;**31**(1):5-17

[12] Camargo-Parra AA, y Rodríguez LNF. Respuestas fenotípicas de Lippia alba y Lippia origanoides (Verbenaceae) a la disponibilidad de agua en el suelo. Acta Biológica Colombiana. 2008;**13**:133-148. Available from: http://www.redalyc.org/articulo.oa?id=319028004011

[13] Do Santos FJ, Lopes JA, Cito GL. Composition and biological activity of essential oils from *Lippia origanoides* H.B.K. Journal of Essential Oil Research. 2004;**16**:504-506. DOI: 10.1080/10412905.2004.9698782

[14] Ramirez LS, Isaza JH, Veloza LA, Stashenko EE. Actividad antibacteriana de aceites esenciales de Lippia origanoides de diferentes orígenes de Colombia. Universidad del Zulia. 2009;**17**:313-321. DOI: 10.1002/ece3.360

[15] Skoula M, Harborne J. The taxonomy and chemistry of Origanum. Genus. In: Spiridon 1 St ed. Oregano, the genera Origanum and Lipia. London and New York: Taylor & Francis; 2002. pp. 127-150. eBook ISBN9780429218972

[16] Lee KW, Everts H, Kappert HJ, Yeom KH, Beynen AC. Dietary carvacrol lowers body weight gain but improves feed conversion in female broiler chickens. Journal of Applied Poultry Research. 2003;**12**:394-399. DOI: 10.1093/japr/12.4.394

[17] Elgayyar M, Draughon FA, Golden DA, Mount JR. Antimicrobial activity of essential oils from plants against selected pathogenic and saprophytic microorganisms. Journal of Food Protection. 2001;**2001**(64):1019-1024. DOI: 10.4315/0362-028x-64.7.1019

[18] Lambert RJW, Skandamis PN, Coote PJ, Nychas GJE. A study of the minimum inhibitory concentration and mode of action of oregano essential oil, thymol and carvacrol. Journal of Applied Microbiology. 2001;**91**:453-462. DOI: 10.1046/j.1365-2672.2001. 01428.x

[19] Ultee A, Kets EOW, Smid EJ. Mechanisms of action of carvacrol on the food-borne pathogen *Bacillus cereus*. Applied and Environmental Microbiology. 1999;**65**:4606-4610. DOI: 10.1128/aem.65.10.4606-4610.1999

[20] Trombetta D, Castelli F, Sarpietro MG, Venuti V, Cristani M, Daniele C, et al. Mechanisms of antibacterial action of three monoterpenes. Antimicrobial Agents and Chemotherapy. 2005;**49**:2474-2478. DOI: 10.1128/AAC.49.6.2474-2478

[21] Vattem DA, Lester CE, DeLeon RC, Jamison BY, Maitin V. Dietary supplementation with two Lamiaceae herbs- (oregano and sage) modulates innate immunity parameters in *Lumbricus terrestris*. Pharmacognosy Research. 2013;**5**:1-9. DOI: 10.4103/0974-8490.105636

[22] Borges Leal ALA, Machado AJT, Bezerra CF, Inácio CES, Rocha JE. Chemical identification and antimicrobial potential of essential oil of *Piper rivinoides* kunth (BETIS-WHITE). Food Chemical Toxicology. 2019;**131**:110559. Available from: http://bit.ly/2LwW7CW

[23] Sarrazin S, Andrade da Silva L, de Assunção AP, Oliveira R, Calao V, da Silva R, et al. Antimicrobial and seasonal evaluation of the carvacrol chemotype oil from *Lippia origanoides* Kunth. Molecules. 2015;**20**:1860-1871. DOI: 10.3390/molecules20021860

[24] Marshall BM, Levy SB. Food animals and antimicrobials: Impacts on human health. Clinical Microbiology Reviews. 2011;**24**:718-733. DOI: 10.1128/CMR.00002-11

[25] Salaheen S, Kim S-W, Haley BJ, VanKessel JAS, Biswas D. Alternative growth promoters modulate broiler gut microbiome and enhance body weight gain. Frontiers in Microbiology. 2017;**8**:2088. DOI: 10.3389/fmicb.2017.02088

[26] Valenzuela-Grijalva N, Pinelli-Saavedra A, Muhlia-Almazan A, Domınguez-Díaz D, González-Ríos H. Dietaryinclusioneffectsofphytochemicals as growth promoters in animal production. Journal of Animal Science and Technology. 2017;**59**:1-11. DOI: 10.1186/s40781-017-0133-9

[27] Krishan G, Narang A. Use of essential oils in poultry nutrition: A new approach. Journal of Advanced Veterinary and Animal Research. 2014;**1**:156-162. DOI: 10.5455/javar.2014.a36

[28] Mehdi Y, Letourneau-Montminy M-P, Gaucher M-L, Chorfi Y, Suresh G, Rouissid T, et al. Use of antibiotics in broiler production: Global impacts and alternatives. Animal Nutrition. 2018;**4**:170-178. DOI: 10.1016/j.aninu.2018.03.002

[29] Zhai H, Liua H, Wanga S, Wua J, Kluente A-M. Potential of essential oils for poultry and pigs. Animal Nutrition. 2018;**4**:179-186. DOI: 10.1016/j.aninu.2018.01.005

[30] Betancourt L, Hume M, Rodriguez F, Nisbet D, Sohail MU, Afanador-Tellez G. Effects of Colombian oregano essential oil (*Lippia origanoides* Kunth) and Eimeria species on broiler production and cecal microbiota. Poultry Science. 2019;**98**(10):4777-4786. DOI: 10.3382/ps/pez193

[31] Dong R, Fan Q, Fouad AM, Sun Y, Huang S, Aiji W, et al. Effects of dietary oregano essential oil supplementation on growth performance, intestinal antioxidative capacity, immunity, and intestinal microbiota in yellow-feathered chickens. Journal of Animal Science. 2021;**99**:1-11

[32] Amer SA, Tolba SA, AlSadek DMM, Abdel Fattah DM, Hassan AM, Metwally AE. Effect of supplemental glycerol monolaurate and oregano essential oil blend on the growth performance, intestinal morphology, and amino acid digestibility of broiler chickens. BMC Veterinary Research. 2021;**17**:2-12

[33] Betancourt L, Rodriguez F, Phandanouvong V, Ariza-Nieto C, Hume M, Nisbet D, et al. Effect of Origanum chemotypes on broiler intestinal bacteria. Poultry Science. 2014;**93**:2526-2535. DOI: 10.3382/ps.2014-03944

[34] Johnson AM, Anderson G, Arguelles-Ramos M, Ali AAB. Effect of dietary essential oil of oregano on performance parameters, gastrointestinal traits, lood lipid profile, and antioxidant capacity of laying hens during the pullet phase. Frontiers in Animal Science. 2022;**3**:1072712. DOI: 10.3389/fanim.2022.1072712

[35] Oviedo-Rondón EO, Hume ME, Hernández C, Clemente-Hernández S. Intestinal microbial ecology of broilers vaccinated and challenged with mixed Eimeria species and supplemented with essential oil blends. Poultry Science. 2006;**85**:854-860. DOI: 10.1093/ps/85.12.2106

[36] Mohiti-Asli M, Ghanaatparast-Rashti M. Dietary oregano essential

oil alleviates experimentally induced coccidiosis in broilers. Preventive Veterinary Medicine. 2015;**120**:195-202. DOI: 10.1016/j.prevetmed.2015.03.014

[37] Danzeisen JL, Kim HB, Isaacson RE, Zheng JT, Johnson TJ. Modulations of the chicken cecal microbiome and metagenome in response to anticoccidial and growth promoter treatment. PLoS One. 2011;**2011**(6):e27949. DOI: 10.1371/journal.pone.0027949

[38] Hume ME, Barbosa NA, Dowd SE, Sakomura NK, Nalian AG, Martynova-Van Kley A, et al. Use of pyrosequencing and denaturing gradient gel electrophoresis to examine the effects of probiotics and essential oil blends on digestive microflora in broilers under mixed Eimeria infection. Foodborne Pathogens and Disease. 2011;**8**:1159-1167. DOI: 10.1089/fpd.2011.0863

[39] Hou Q, Kwok LY, Zheng Y, Wang L, Guo Z, Zhang J, et al. Differential fecal microbiota are retained in broiler chicken lines divergently selected for fatness traits. Scientific Reports. 2016;**2016**(6):1-10. DOI: 10.3920/BM2014.0063

[40] Caricilli A, Saad M. The role of gut microbiota on insulin resistance. Nutrients. 2013;**5**:829-851. DOI: 10.3390/nu5030829

[41] Betancourt LL, Ariza CJ, Afanador G. Effects of supplementation with oregano essential oil on ileal digestibility, intestinal histomorphology, and performance of broiler chickens. Revista Colombiana de Ciencias Pecuarias. 2012;**25**:240-251. DOI: 10.17533/udea.rccp.324751

[42] Jia L, Wu J, Lei Y, Kong F, Zhang R, Sun J, et al. Oregano essential oils mediated intestinal microbiota and metabolites and improved growth performance and intestinal barrier function in sheep. Frontiers in Immunology. 2022;**13**:908015. DOI: 10.3389/fimmu.2022.908015

[43] Nusrat A, Turner JR, Madara JL. Molecular physiology and pathophysiology of tight junctions IV. Regulation of tight junctions by extracellular stimuli: Nutrients, cytokines, and immune cells. American Journal of Physiology. Gastrointestinal and Liver Physiology. 2000;**279**:G851-G857. DOI: 10.1152/ajpgi.2000.279.5.G851

[44] Chen J, Tellez G, Richards JD, Escobar J. Identification of potential biomarkers for gut barrier failure in broiler chickens. Frontiers in Veterinary Science. 2015;**2**:14. DOI: 10.3389/fvets.2015.00014

[45] Stashenko EE, Martínez JR, Durán DC, Yuri Córdoba+Deyanira Caballero. Estudio comparativo de la composición química y la actividad antioxidante de los aceites esenciales de algunas plantas del género Lippia (Verbenaceae) cultivadas en Colombia. Revista de la Academia de Ciencias Exactas, físicas y naturales. 2014;**38**:11-28. DOI: 10.18257/raccefyn.156

[46] Oscar A, Pantoja D, Santacruz L, Hurtado AM. Antioxidant activity of essential oils of oregano (Lippia Origanoides H.B.K) grown in alto Patia. Biotecnología en el Sector Agropecuario y Agroindustrial. 2012;**10**:79-86

[47] Mendez-Zamora G, García Macías JA, Durán-Meléndez LA, Silva-Vázquez R. Aceite esencial de orégano (Lippia berlandieri Schauer) en variables de calidad de la canal de pollo. Ecosistemas y Recursos Agropecuarios. 2015;**2**:41-51

[48] Oliveira R, Leitão GG, Fernandes PD, Leitão SG. Ethno pharmacological studies

of Lippia origanoides. Revista Brasileira de Farmacognosia. 2014;**24**:206-214

[49] Zhang R, Wang XW, Liu LL, Cao YC, Zhu H. Dietary oregano essential oil improved the immune response, activity of digestive enzymes, and intestinal microbiota of the koi carp, Cyprinus carpio. Aquaculture. 2020;**518**:734781. DOI: 10.1016/j.aquaculture.2019.734781

Section 4

Essential Oil Applications

Chapter 8

Essential Oil Extraction Process

Zoubeida Souiy

Abstract

Essential oils can be extracted using various methods. Process choice significantly impacts yield and quality, leading to the development of processes aiming for maximum essential oil (EO) yields in a chemical state close to their native structure. In this chapter, various extraction techniques, including conventional ones and their intensification, are discussed along with their respective pros and cons. Additionally, new eco-friendly extraction methods have been introduced to enhance the conventional production of essential oils. The most traditional, straightforward, and widely utilized extraction techniques are hydrodistillation and steam extraction. In actuality, steam extraction techniques are used to extract 93% of all essential oils. Other common extraction techniques include enfleurage (particularly used with roses), cold pressing (just for citrus peel), and organic solvent extraction. The low yield, loss of volatile chemicals, lengthy extraction durations, and hazardous solvent residues of these procedures are its drawbacks. Microwave-assisted extraction and supercritical fluid extraction are two of the latest essential oil extraction techniques that have received considerable interest.

Keywords: essential oils, hydrodistillation, steam extraction, microwave-assisted extraction, supercritical fluid extraction

1. Introduction

Essential oil (EO) is a secondary metabolite synthesized by medicinal and aromatic plants. It represents less than 5% of the total plant composition. Over 3000 types of EO have been identified, but only 300 were found to be economical [1, 2].

EO is volatile, generally colorless, and liquid at room temperature. It is highly soluble in organic solvents, alcohol, and fixed oils but sparingly soluble in water. It has very high optical activity, a variable refractive index, and sometimes a distinctive taste. In addition, essential oils have a characteristic odor, that is responsible for the fragrance specific to the aromatic plant. Chemically, EO components can be divided into terpene compounds and aromatic compounds. It is a mixture of bi-active chemical components such as terpenoids, terpenes, and phenolic compounds. They are made up of terpene compounds, acids, alcohols, esters, aldehydes, ketone epoxides, sulfides, and amines [3, 4].

They are synthesized by the majority of plant organs, in particular flowers, buds, leaves, seeds, stems, and fruits. These EOs can be stored in epidermal cells, cavities, and the secretory cells of glandular trichomes.

EOs are known for their biological activity, notably their antioxidant, antifungal, antimicrobial, antiviral, antiparasitic, antimycotic, and insecticidal properties [5, 6].

Several extraction techniques are used; Hydrodistillation and steam extraction are the oldest, simplest, and most commonly used methods. Other extraction methods can also be used: cold pressing, especially applied to rose.

The EO extraction method generally depends on the botanical material used. It is one of the main factors determining EO quality. An inappropriate extraction procedure can damage or alter the chemical composition of EO. This results in a loss of bio-activity and natural characteristics. In the most serious cases, this is accompanied by discoloration, an unpleasant odor or flavor, and physical changes such as increased viscosity [7].

The principle of EO extraction is relatively straightforward. However, the process chosen can have a significant effect on the yield and quality of the distillate obtained [8]. Various processes have therefore been developed to obtain maximum yields of EO with a chemical state as close as possible to their native structure.

According to the European Pharmacopeia, an essential oil can be obtained by steam distillation, distillation, or a mechanical process. Other processes include supercritical CO_2 extraction, volatile organic solvent extraction, microwave extraction, and ultrasonic extraction. The aim of this chapter is to present an overview of the various extraction methods.

2. Location and yield of essential oil

Plants have the natural ability to produce volatile compounds in trace amounts. However, only a small percentage of plant species, around 10%, are considered "aromatic". This property of accumulating essential oils is found in specific plant families distributed throughout the plant kingdom, including Pinacea (pine and fir), Cupressaceae (cedarwood), and angiosperms. The most significant families are dicotyledons such as Asteracea (chamomile), Apiaceae (coriander), Geraniaceae (geranium), Lamiaceae (mint), Illiciaceae (anise), Lauraceae (cinnamon), Rosacea (rose), Sandatalacea (sandalwood), Myrtaceae (eucalyptus), Myristicaceae (walnut), Oleacea (jasmine), and Rutacea (lemon). Monocotyledons are mainly represented by the families Zingiberaceae (ginger), and Poacea (vetiver) [9, 10].

EOs are natural secretions produced by cells and found in plant parts such as flowers (rose), leaves (lemongrass), flowering tops (lavender), bark (cinnamon), roots (iris), bulbs (garlic), fruits (vanilla), seeds (nutmeg), or rhizomes (ginger). Essential oils are extracted from specific parts of plants, such as sage or lavender. The most concentrated or secretory parts of the plant are harvested at the optimum yield period, which varies depending on the plant. For example, mints are harvested before flowering, lavenders during flowering, and seed plants after flowering or after morning dew for fragile flowers. It is important to note that plant growth conditions can also affect yield and essential oil content. The collection period and drying methods can also impact the yield. Therefore, it is crucial to choose the right harvesting time and drying and extraction methods to obtain the maximum yield and quality of essential oils [8].

This text reviews both traditional and "green" extraction techniques, comparing their performance with conventional methods and emphasizing the benefits of "green" technology in plant extraction research.

3. Conventional extraction methods

Conventional extraction methods can have some drawbacks, such as the degradation of unsaturated compounds and loss of certain components. It is great to hear that there are ongoing efforts to improve and optimize extraction techniques and that these techniques are carefully chosen based on the plant organ and desired product quality. It is also important to note that the analytical composition of EOs can vary depending on the extraction technique used and that factors such as distillation duration, temperature, operating pressure, and raw plant material quality can all influence EO yield [11].

3.1 Steam extraction

Steam extraction (**Figure 1**) is a widely used and official method for extracting essential oils from plants. This method accounts for 93% of essential oil extractions and can take anywhere from 1 to 10 hours depending on factors such as extraction time, temperature, pressure, and type of material [12].

In this extraction system, plant material is exposed to a stream of steam without prior maceration. The heat applied breaks down the cells of the plant material, releasing the essential oil. The steam, saturated with volatile compounds, is then condensed, and the essential oil is recovered by decanting the water/oil mixture [13, 14].

One of the advantages of steam extraction is that the absence of direct contact between water and plant material, and then between water and aromatic molecules, prevents hydrolysis or degradation of essential oil [15]. The "head" fractions, which contain the most volatile molecules, can be collected in as little as half an hour, with 95% of the volatile molecules being collected [16].

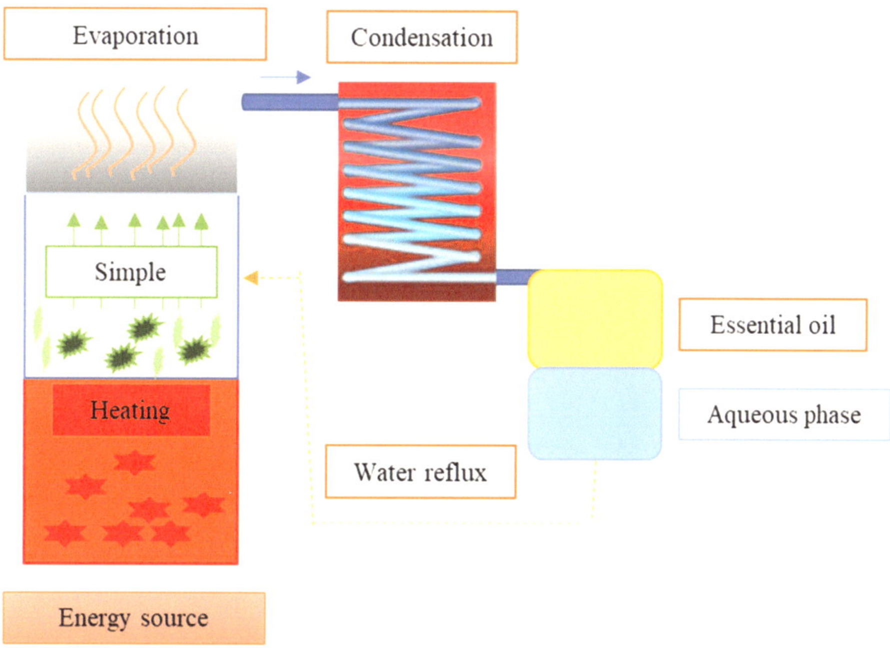

Figure 1.
A schematic representation of steam extraction of essential oils.

The technique works by ensuring that the combined vapor pressure equals the ambient pressure at about 100°C, allowing volatile components with boiling points ranging from 150 to 300°C to be evaporated at a temperature close to that of water. It is also interesting to note that this technique can be carried out under pressure depending on the extraction difficulty of the EOs [17].

3.2 Hydrodistillation

Hydrodistillation (HD) (**Figure 2**) is a standard EO extraction method. It enables the extraction of water-insoluble natural products with a high boiling point. The process involves complete immersion of the plant material in water, followed by boiling. This operation is generally carried out under atmospheric pressure. The steam formed is condensed by the refrigeration system at a water flow rate.

This method protects the extracted oils from overheating. The advantage of this technique is that the required material can be distilled at temperatures below 100°C.

Distillation may seem like a simple process for extracting essential oils, but it comes with several drawbacks. In developed countries, its use has become outdated due to the overheating of plant material and the production of burned-smelling oils. However, this method is still effective for powders and hard materials. It is important to note that exposure to boiling water for extended periods can cause weathering reactions and hydrolysis of esters into alcohols and acids, which can have serious consequences for oils with high ester levels. Rectification is often necessary to remove unwanted impurities or constituents responsible for unacceptable odor. Distillation time varies depending on the type of plant material, with woody plant organs requiring longer distillation times than herbaceous plants [18].

3.3 Hydrodiffusion

Hydrodiffusion is another method conventional method for extracting essential oils from plant materials. It involves the use of steam and water to extract the oils.

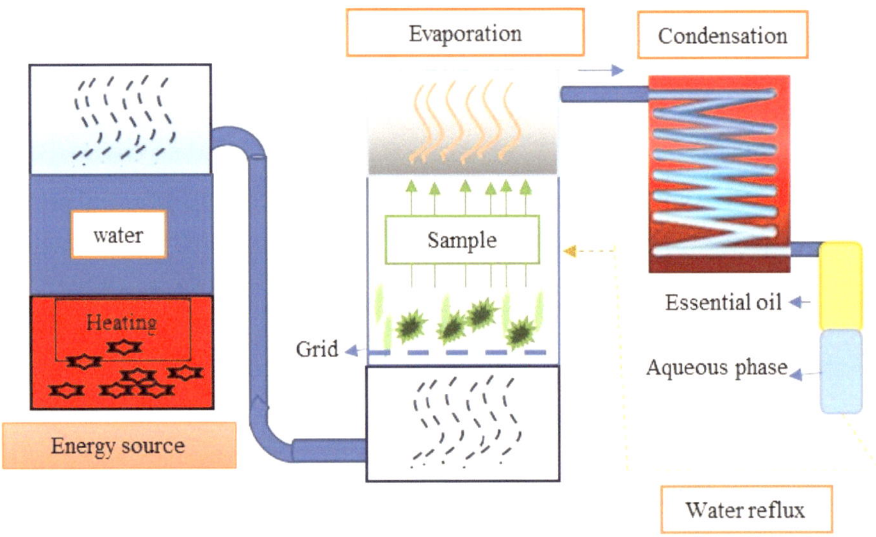

Figure 2.
A schematic representation of hydrodistillation of essential oils.

The plant material is placed on a grid above water in a distillation vessel, and steam is injected into the bottom of the vessel. The steam then passes through the plant material, carrying the essential oils with it. The steam and oil mixture then condenses on a cooled surface, with the oil and water separating into two layers. This method is particularly useful for extracting essential oils from delicate plant materials, as it uses lower temperatures and less pressure than other methods [19, 20].

3.4 Cold pressing

One of the oldest extraction methods for essential oils from citrus peels such as lemon, orange, bergamot, and grapefruit is cold pressing. This technique mechanically tears the peels by simply pressing them to extract the volatile essences contained in the citrus pericarps. Until the early twentieth century, cold-pressed citrus oils were produced manually. The process produces an aqueous emulsion, which is then centrifuged to separate the essential oil. This method is preferred for citrus peel essential oil extraction because it avoids thermal alteration of the aldehydes. This process results in the production of an aqueous emulsion, which is then centrifuged to separate the EO [21].

3.5 Enfleurage

Enfleurage is another conventional extraction method that dates back to antiquity. It is based on the affinity of fragrances for fats, and concerns plants that retain their fragrance after being picked (such as jasmine or tuberose). The flowers are spread out on frames coated with odorless grease. The flowers' fragrance is absorbed by the grease until saturation. The flowers are changed regularly (e.g., every 24 hours for jasmine). When the fat is saturated by the flowers, the operation is complete. Saturation can last up to a month. The resulting pomade is then melted. After decanting, the mixture is cold-treated with alcohol. The alcohol draws out the fragrance on its own, without taking on the fats. This extraction technique is virtually dying out due to its high cost, and the extracted oils have no applications in the food industry [14, 22, 23].

3.6 Organic solvent extraction

Solvent extraction is commonly employed to extract EOs that exhibit thermal labile properties, such as those extracted from flowers. The plant material is placed in a solvent bath. Successive washings charge the solvent with aromatic molecules. After separation by filtration, the emulsion is distilled to extract the EO.

Solvent extraction has been used for fragile or delicate floral materials, which cannot withstand the temperature of distillation. Various solvents, including hexane, acetone, petroleum ether, ethanol, or methanol, can be used for extraction [7].

Solvent extraction is relatively fast and inexpensive. The chosen solvent must be permissible, inert, and stable to heat, light, or oxygen. Its boiling temperature should preferably be low to facilitate elimination.

The produced EO contains a small amount of solvent residue, making it unsuitable for food applications. However, if alcohol is used as the solvent, it is considered "food-grade" and safe for consumption. This method is commonly used in the perfume industry [24].

In practice, the solvent is mixed with the plant material, heated to extract the EO, and then filtered. The filtrate is then concentrated through solvent evaporation. It is later mixed with pure alcohol to extract the oil and distilled at low temperature.

However, this method is relatively time-consuming, making the oils more expensive than other methods. Additionally, solvent residues in the final product can cause allergies, toxicity, and affect the immune system [25].

The limited use of this extraction method is justified by its cost, toxicity and safety issues, and environmental protection regulations. However, HE yields are generally higher than with distillation. What is more, this technique avoids the hydrolyzing action of water vapor.

4. New "green" extraction methods

4.1 Microwave-assisted extraction

Since 1986, microwave energy has been widely used in chemistry laboratories. Researchers have studied the potential of this unconventional energy source for synthetic, analytical, and processing applications. Currently, there are over 3000 articles documenting the use of dielectric heating in synthesis and over 1000 articles documenting its use in extraction.

Microwave-assisted extraction is a revolutionary technology that has garnered a lot of interest. It has a distinctive friction-based heating mechanism. It is inexpensive, and performs well under atmospheric conditions.

Microwave-assisted extraction achieves higher extraction yields, shorter extraction times, and improved selectivity as compared to traditional extraction techniques. This process is also less complicated and expensive than supercritical fluid extraction. However, it usually requires for more organic solvent, which makes it less environmentally friendly [26].

Recent methods of microwave-assisted extraction include microwave-assisted vacuum hydrodistillation, compressed air distillation, and microwave-assisted accelerated steam distillation [27].

4.1.1 Dielectric heating and fundamentals of microwave extraction

Microwave irradiation utilizes a specific electromagnetic field frequency, similar to activated photochemical reactions. The frequency range is vast, extending from 300 MHz to 300 GHz, but only certain frequencies are authorized for industrial, scientific, and medical use. These include frequencies of 0.915 and 2.45 GHz. The magnetron, found in domestic and laboratory microwave furnaces, is a typical microwave generator for such frequencies. Industrial magnetrons can reach powers of several tens of kilowatts, while laboratory devices generally have powers of less than 1 kW. Solid-state generators have recently been introduced, which narrow the microwave generator's emission band, allowing the user to vary the system's frequency within the range of authorized industrial, scientific, and medical frequencies. This variation can play a crucial role in chemical synthesis, particularly with regard to selectivity and efficiency. However, solid-state generators operating at 2.45 GHz typically have a power rating of 100 W, which is also frequently used in medical applications [28].

Microwave-assisted extraction (MAE) is a process that removes solutes from a solid matrix into a solvent. The process involves complex phenomena such as heat transfer electromagnetic transfer, mass transfer, and momentum transfer [29].

4.1.2 Microwave solvent-assisted extraction

Microwave solvent-assisted extraction (**Figure 3**) have revolutionized the field of bioactive compound extraction. This technique has significantly reduced extraction times, minimized organic solvent consumption, and resulted in energy and cost savings [30].

Moreover, microwave solvent-assisted extraction is an environmentally friendly and sustainable method that contributes to the development of "green" procedures.

A new and efficient method for extracting essential oils from *Angelica sinensis* root has been developed using a microwave-assisted deep eutectic natural solvent (NADES). The study found that NADES based on choline chloride and citric acid were more effective in extracting essential oils [31].

4.1.3 Compressed air microwave distillation (CAMD)

This method (**Figure 4**) uses the principle of steam entrainment, with compressed air instead of steam, to extract the essential oil. The extraction process consists of a compressor, a microwave oven and a refrigeration system. Compressed air is injected into the reactor, where the matrix is heated by microwaves and immersed in water. The steam, saturated with volatile molecules, is directed to a recovery container located outside the microwave oven and cooled by a refrigeration system. In just a few minutes, the water and aromatic molecules are condensed and recovered [30, 32].

A similar method using a condenser to cool the extraction gas (temperatures ranging from −20 to −15°C) has also been patented [33]. This extraction method is environmentally friendly, as no organic solvents or artificial chemical compounds are added.

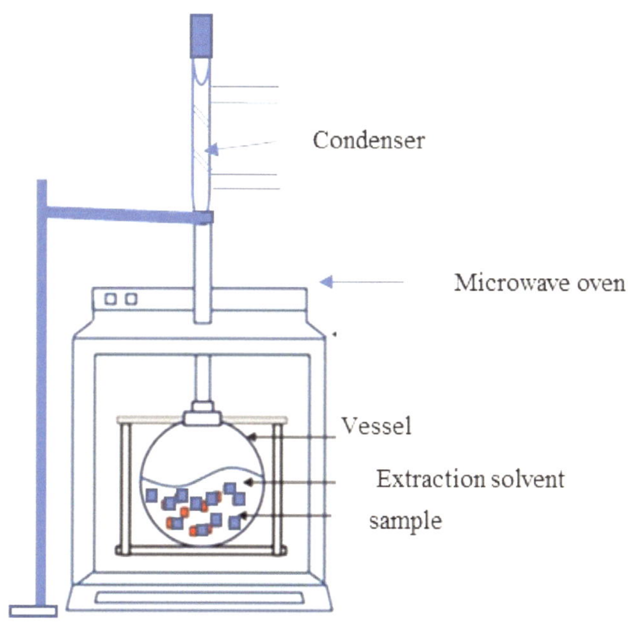

Figure 3.
Experimental set-up for microwave solvent-assisted extraction.

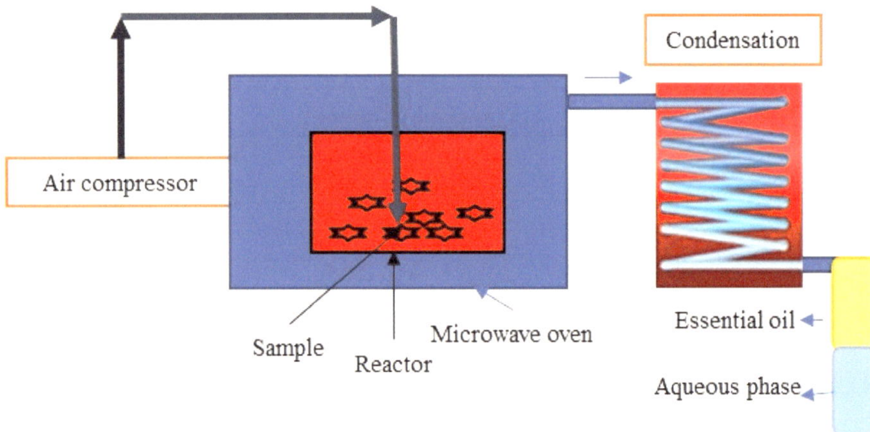

Figure 4.
A schematic representation compressed air microwave distillation.

4.1.4 Microwave hydrodistillation (MWHD)

The MWHD (**Figure 5**) was developed by Stashenko et al., in 2004. It is based on the classic hydrodistillation principle. The process consists of a hydrodistillation unit placed inside a domestic microwave oven with a side port, through which an external glass condenser is connected to the round filter containing the matrix and water [34].

Microwave hydrodistillation is a widely used technique for extracting essential oils from various aromatic plants and spices, with examples such as *Thymus vulgaris* L., *Zataria multiflora* Boiss., and *Satureja montana*.

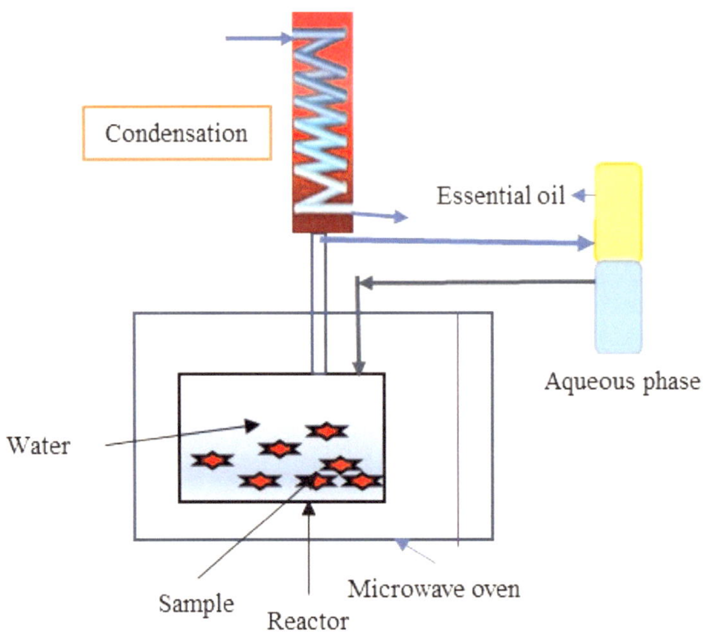

Figure 5.
Microwave hydrodistillation.

An improved version of this technique was developed in 2007, which involves introducing a microwave coaxial antenna insulated inside a glass flask containing the matrix and water [35]. This in situ microwave heating offers advantages in terms of time and energy savings and can be useful for industrial applications.

Microwave Steam Distillation (MSD) (**Figure 6**) is another innovative technique that was developed. It is based on the conventional steam distillation principle and has been successfully used for the extraction of essential oil from Lavender flowers [30].

4.1.5 Solvent-free microwave extraction (SFME)

This is one of the most recent techniques for the microwave-assisted extraction of essential oils, without solvents and using water at atmospheric pressure. The SFME process consists mainly of four parts: a reactor where the matrix to be treated is placed, a microwave oven, a cooling system, and an essential oil container where the oil is collected (**Figure 7**).

The process is based on a relatively simple principle, described as microwave-assisted dry distillation; the fresh matrix is placed in a microwave reactor without the addition of water or organic solvent. Heating the raw material with water breaks down the glands containing the essential oil. This phase releases the essential oil, which is then carried away by the steam produced by the water in the matrix. A cooling system located outside the microwave oven enables continuous condensation of the distillate, composed of water and essential oil, and the return of excess water

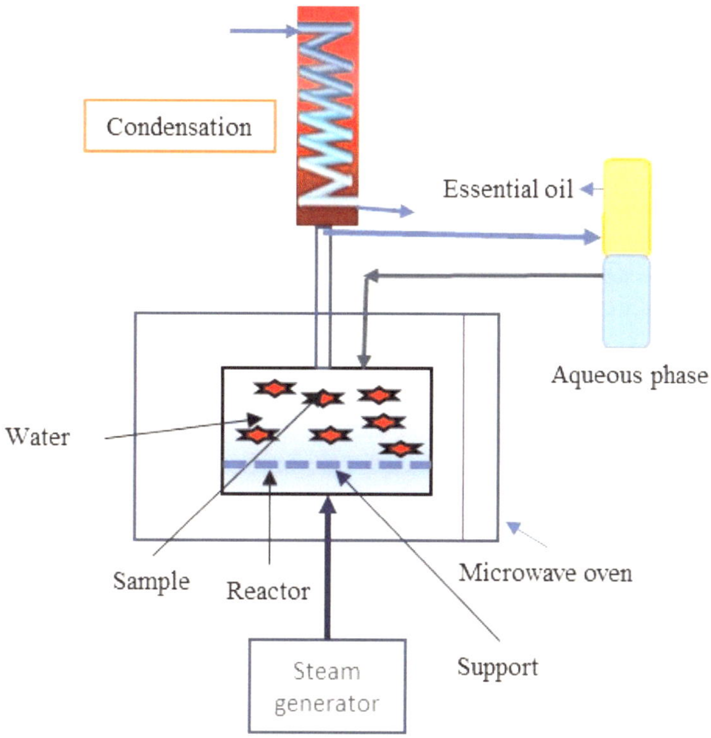

Figure 6.
Microwave steam distillation (MSD).

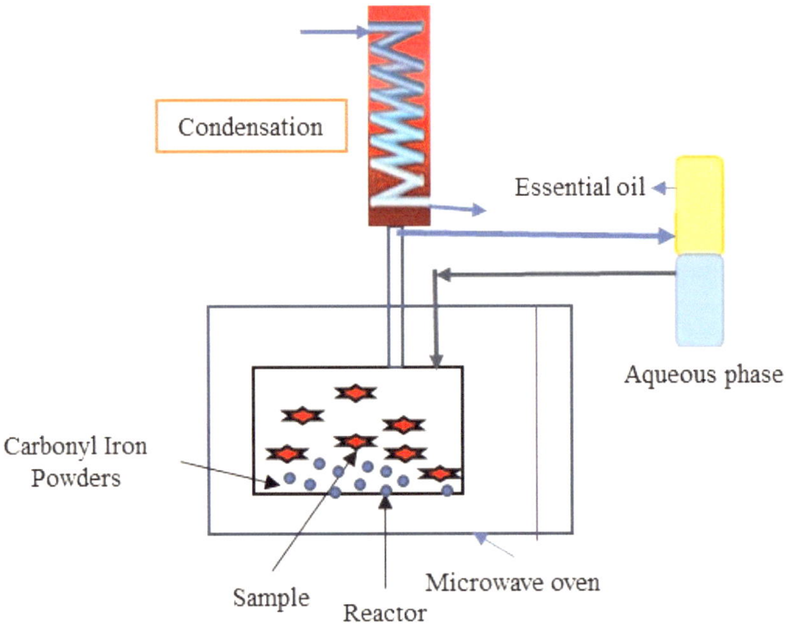

Figure 7.
Improved solvent-free microwave extraction (improved SFME).

to the bottle, thus maintaining the appropriate moisture content of the matrix. For example, Milestone's "DryDist" laboratory microwave oven makes it easy and efficient to extract high-quality essential oils.

Wang et al. in 2006 proposed an improved SFME extraction method. The method is based on the addition and mixing of carbonylated iron powder with the dry matrix. Spherical particles of carbonylated iron are capable of absorbing part of the energy emitted by microwaves and returning it to the medium in the form of heat. In this way, the matrix can be heated by simple conduction without any auxiliary energy. Various types of materials such as activated carbon, graphite powders, and ionic liquid (1-hexyl-3-methylimidazolium hexafluorophosphate) can absorb microwave radiation [36].

4.1.6 Microwave hydrodiffusion and gravity (MHG)

The Microwave Hydrodiffusion and Gravity (MHG) process was invented by a team of researchers led by Dr. Farid Chemat at the University of Avignon in France [37]. The team developed the MHG process as an alternative to traditional methods of essential oil extraction, which can be time-consuming and require large amounts of energy. The MHG process was first introduced in 2004 and has since gained popularity in the essential oil industry due to its efficiency and effectiveness.

The process of Microwave Hydrodiffusion and Gravity (MHG) involves the following steps:

- Preparation: The plant material, such as herbs or flowers, is first cleaned and dried to remove any impurities.

- Loading: The dried plant material is placed in a vessel that is suitable for microwave heating, such as a glass container or a microwave-safe bag.

- Microwave Heating: The vessel containing the plant material is exposed to microwave radiation. The microwaves generate heat, causing the essential oil compounds within the plant material to vaporize.

- Hydrodiffusion: As the plant material is heated, water molecules present in the plant cells also vaporize. This creates a hydrodiffusion effect, where the vaporized water carries the essential oil compounds with it.

- Condensation: The vapor containing the essential oil compounds and water is then cooled down, causing it to condense. The condensation occurs in a separate container or condenser unit.

- Separation: The condensed mixture of essential oil and water is then separated. This can be done using techniques such as decantation or using a separating funnel.

- Collection: The essential oil, which is lighter than water, floats on top and can be collected from the separated mixture.

- Analysis and Storage: The collected essential oil can be analyzed for quality and stored in suitable containers to preserve its aroma and therapeutic properties.

4.2 Supercritical fluid extraction

Supercritical fluid extraction (SFE) is a separation technique that utilizes supercritical fluids as the extracting solvent. A supercritical fluid is a substance that is above its critical temperature and pressure, which results in unique properties that make it an effective solvent for extraction.

The principle of SFE is based on the fact that the solubility of a substance in a supercritical fluid increases with pressure, while the density of the fluid increases with pressure and temperature. By adjusting the temperature and pressure, the solubility of the substance can be controlled and optimized for extraction.

In SFE, the supercritical fluid is pumped into a vessel containing the sample to be extracted. As the fluid passes through the sample, it dissolves the target compounds, which are then carried out of the vessel and into a collection vessel by depressurization or by lowering the temperature. The extracted compounds can then be separated from the supercritical fluid by condensation or by other means.

SFE has several advantages over traditional extraction methods, including reduced solvent use, shorter extraction times, and higher yields of target compounds.

Supercritical fluid extraction (SFE) can be performed in a variety of ways: batch, semi-batch, or continuous. Plant material is placed in a container and supercritical fluid is added at a specific flow rate until the appropriate extraction conditions are reached. Compared with conventional solvent extraction methods, supercritical fluid extraction offers several advantages, including a lower temperature suitable for thermosensitive compounds and a solvation power that can be controlled by modifying pressure and/or temperature, enabling high selectivity. Supercritical fluids are more effective than liquid solvents in penetrating porous materials and extracting compounds, resulting in faster extraction and a more environmentally friendly process. CO_2 and small amounts of organic solvents can be used as nontoxic fluids, and this method can be used on an industrial scale [38].

However, high pressures should be avoided when extracting essential oils to prevent the extraction of undesirable compounds.

To ensure the success of EFS, various factors need to be taken into account, such as sample type, preparation, fluid type, delivery method, and extraction conditions. CO_2 is commonly used due to its low critical temperature, cost-effectiveness, nontoxicity, absence of odor and taste, and ease of disposal. Adjusting the process conditions makes it possible to selectively extract the desired components. Compared with steam distillation, EFS has shorter extraction times, lower energy costs, and greater selectivity. The EFS method also makes it easier to manipulate oil composition by modifying extraction parameters [39].

5. Conclusion

In conclusion, there are multiple methods for extracting essential oils, and the process chosen can greatly affect the amount and quality of the oil produced. To maximize yields and maintain the natural structure of the oils, extraction processes have been developed. This chapter explores different extraction techniques, both conventional and intensified, highlighting their advantages and disadvantages. It is improved that new techniques have been proven to produce higher quality extracts in a shorter time compared to traditional techniques. However, regulatory standards do not list these extracts derived from innovative techniques as essential oils due to the narrow definition of essential oils based solely on conventional extraction methods. Furthermore, new environmentally friendly methods have been introduced to improve traditional essential oil production. Therefore, it is becoming increasingly crucial to modify or re-establish industry standards to encompass a broader range of extraction techniques.

Author details

Zoubeida Souiy[1,2]

1 Biochemistry Laboratory, LR12ES05 “Nutrition-Functional Foods and Vascular Health”, Faculty of Medicine, University of Monastir, Tunisia

2 Higher Institute of Technology Studies ISET (Ksar Hellal), Tunisia

*Address all correspondence to: souiyzoubeida@gmail.com

References

[1] Sil A et al. Essential oils: A boon towards eco-friendly management of phytopathogenic fungi. Journal of Entomology and Zoology Studies. 2020;**8**:1884-1891

[2] Nik NNA et al. Essential Oils Composition and Bioactivities of Most Prevalent Species of Genus Piperin Malaysia: A Scope Review. Vol. 1002 art. La rivista italiana delle sostanze grasse; 2023. Available from: en.innovhub-ssi.it

[3] da Silva WMF et al. Basil essential oil: Methods of extraction, chemical composition, biological activities, and food applications. Food and Bioprocess Technology. 2022;**15**(1):1-27

[4] Sadgrove NJ, Padilla-González GF, Phumthum M. Fundamental chemistry of essential oils and volatile organic compounds, methods of analysis and authentication. Plants. 2022;**11**(6):789

[5] Hou T et al. Essential oils and its antibacterial, antifungal and anti-oxidant activity applications: A review. Food Bioscience. 2022;**47**:101716

[6] Souiy Z et al. Application of simplex-centroid design methodologies to optimize the anti-bacterial and anti-candidal activity of the mixture of menthapulegium, pituranthos chloranthus and thymus algeriensis essential oils. Medicinal & Aromatic Plants. 2021;**10**:365

[7] Tongnuanchan P, Benjakul S. Essential oils: Extraction, bioactivities, and their uses for food preservation. Journal of Food Science. 2014;**79**(7):R1231-R1249

[8] Souiy Z et al. Optimization of extraction yield, chemical composition, antioxidant, and antimicrobial activities of Thymus algeriensis Boiss. & Reut essential oils. Biocatalysis and Agricultural Biotechnology. 2023;**47**:102593

[9] Spichiger R-E, Figeat-Hug M, Jeanmonod D. Botanique systématique des plantes à fleurs: Une approche phylogénétique nouvelle des angiospermes des régions tempérées et tropicales. Lausanne, Suisse: PPUR Presses Polytechniques; 2002

[10] Roux D, Catier O. Botanique, Pharmacognosie, Phytothérapie. 3e éd ed. Rueil-Malmaison: Wolters Kluwer France; 2007

[11] Prado JM et al. Conventional extraction. In: Food Waste Recovery. Amsterdam, Netherlands: Elsevier; 2021. pp. 109-127

[12] Nourelhouda BAARS. Etude Comparative des méthodes d'extraction des huiles Essentielle dans les plantes. Algeria: Universite Mohamed Boudiaf-M'SILA; 2021

[13] Dugo G, Di Giacomo A, editors. Citrus: The Genus Citrus. 1st ed. London: CRC Press; 2002. DOI: 10.1201/9780203216613

[14] Souiy Z et al. Nutritional, physical, microbial, and sensory characteristics of gluten-and sugar-free cereal bar enriched with spirulina and flavored with neroli essential oil. LWT. 2022;**169**:113955

[15] Raaman N. Phytochemical Techniques. Delhi, India: New India Publishing; 2006

[16] Gavahian M, Chu Y-H. Ohmic accelerated steam distillation of essential oil from lavender in comparison with conventional steam distillation.

Innovative Food Science & Emerging Technologies. 2018;**50**:34-41

[17] Li Y, Fabiano-Tixier AS, Chemat F. Essential oils: From conventional to green extraction. In: Essential Oils as Reagents in Green Chemistry. SpringerBriefs in Molecular Science. Cham: Springer; 2014. DOI: 10.1007/978-3-319-08449-7_2

[18] Katekar VP, Rao AB, Sardeshpande VR. Review of the rose essential oil extraction by hydrodistillation: An investigation for the optimum operating condition for maximum yield. Sustainable Chemistry and Pharmacy. 2022;**29**:100783

[19] Murti Y et al. Innovative methods for extraction of essential oils from medicinal plants. International Journal of Secondary Metabolite. 2023;**10**(2):190-230

[20] Gupta A et al. Essential oils: Chemical composition and methods of extraction. In: Essential Oils: Extraction Methods and Applications. Hoboken, New Jersey, USA: John Wiley & Sons; 2023. pp. 871-889

[21] Bousbia N. Extraction des huiles essentielles riches en anti-oxydants à partir de produits naturels et de co-produits agroalimentaires. France: Université d'Avignon; 2011

[22] Stratakos AC, Koidis A. Methods for extracting essential oils. In: Essential Oils in Food Preservation, Flavor and Safety. Amsterdam: Elsevier; 2016. pp. 31-38

[23] Souiy Z et al. The use of D-optimal mixture design in optimizing formulation of a nutraceutical hard candy. International Journal of Food Science. 2023;**2023**:7510452

[24] Hanif MA et al. Essential oils. In: Essential Oil Research. Berlin, Germany: Springer; 2019. pp. 3-17

[25] Ferhat MA, Meklati BY, Chemat F. Comparison of different isolation methods of essential oil from citrus fruits: Cold pressing, hydrodistillation and microwave 'dry'distillation. Flavour and Fragrance Journal. 2007;**22**(6):494-504

[26] Zhang H-F, Yang X-H, Wang Y. Microwave assisted extraction of secondary metabolites from plants: Current status and future directions. Trends in Food Science & Technology. 2011;**22**(12):672-688

[27] Sahraoui N et al. Improved microwave steam distillation apparatus for isolation of essential oils: Comparison with conventional steam distillation. Journal of Chromatography A. 2008;**1210**(2):229-233

[28] Leonelli C, Veronesi P, Cravotto G. Microwave-assisted extraction: An introduction to dielectric heating. In: Microwave-Assisted Extraction for Bioactive Compounds: Theory and Practice. Berlin, Germany: Springer; 2012. pp. 1-14

[29] Veggi PC, Martinez J, Meireles MAA. Fundamentals of microwave extraction. In: Microwave-Assisted Extraction for Bioactive Compounds: Theory and Practice. Berlin, Germany: Springer; 2012. pp. 15-52

[30] Chemat F, Abert-Vian M, Fernandez X. Microwave-assisted extraction of essential oils and aromas. In: Microwave-Assisted Extraction for Bioactive Compounds: Theory and Practice. Berlin, Germany: Springer Science & Business Media; 2013. pp. 53-68

[31] Fan Y, Li Q. An efficient extraction method for essential oil from angelica sinensis radix by natural deep eutectic solvents-assisted microwave

hydrodistillation. Sustainable Chemistry and Pharmacy. 2022;**29**:100792

[32] Craveiro A et al. Microwave oven extraction of an essential oil. Flavour and Fragrance Journal. 1989;**4**(1):43-44

[33] Lee B-N. Microwave Apparatus and Method of Extracting Essential Oils, Essence, and Pigments with Gas Flow. California, USA: Google Patents; 2012

[34] tashenko EE, Jaramillo B.E, Martínez JR. Comparison of different extraction methods for the analysis of volatile secondary metabolites of Lippia alba (mill.) NE brown, grown in Colombia, and evaluation of its in vitro antioxidant activity. Journal of Chromatography A. 2004;**1025**(1):93-103

[35] Flamini G et al. Comparison between the conventional method of extraction of essential oil of Laurus nobilis L. and a novel method which uses microwaves applied in situ, without resorting to an oven. Journal of Chromatography A. 2007;**1143**(1-2):36-40

[36] Wang Z et al. Improved solvent-free microwave extraction of essential oil from dried Cuminum cyminum L. and Zanthoxylum bungeanum maxim. Journal of Chromatography A. 2006;**1102**(1-2):11-17

[37] Chemat F et al. Review of alternative solvents for green extraction of food and natural products: Panorama, principles, applications and prospects. Molecules. 2019;**24**(16):3007

[38] Yousefi M et al. Supercritical fluid extraction of essential oils. TrAC Trends in Analytical Chemistry. 2019;**118**:182-193

[39] Souiy Z, Ben Moussa A, Ksibi H. Numerical simulation of heat and mass transfers in a supercritical dissolution column. Chemical Engineering & Technology: Industrial Chemistry-Plant Equipment-Process Engineering-Biotechnology. 2007;**30**(6):715-720

Chapter 9

Application of Essential Oils on Active Packaging Systems

Imen Dridi, Ahmed Landoulsi and Nadia Smirani

Abstract

Millions tons of food waste are annually generated, causing serious environmental problems. Indeed, the degradation of food quality occurs naturally due to its vulnerability to biochemical reactions such as protein degradation, lipid oxidation, and microbiogical attacks. This huge waste mass can be minimized throughout the food supply chain by many methods including increasing the shelf life of products. Therefore, active food packaging, which not only contains and protects food but also interacts with packaged products, is used. Nevertheless, the migration process, which is defined by the transfer of chemical compounds from the food packaging to the food, may cause changes in the food product quality and safety. Active packaging can contain several additives, allowing them to have antibacterial, antioxidant activities, oxygen, carbon dioxide, and ethylene scavengers, carbon dioxide emitters, odor emitters and absorbers, relative humidity regulators antibacterial antioxidants. Essential oils (EOs) are popular for their natural antimicrobial and antioxidant properties that are increasing consumer demand due to the perception of their 'safer' natural origin. The purpose of this chapter is to study the incorporation of EOs in the active packaging formulation.

Keywords: food waste, active packaging, essential oils, incorporation, antibacterial activity, antioxidant activity

1. Introduction

According to FAO (2019), approximately 14% of the world's food production is lost, resulting in a contribution of 8–10% to global greenhouse gas emissions [1]. Thus, food waste is a significant economic, environmental, and social issue. Consequently, ensuring sustainable consumption and production patterns by reducing food losses along production and supply chains is one of the United Nations' 2030 Agenda for Sustainable Development targets [2]. Food lost waste (FLW) can be defined as the mass of food wasted during food chains. FLW can occur during production, postharvest, and processing stages.

One of the solutions to reduce FLW is the extension of food shelf life by innovative packaging technologies. Food packaging has a critical role in the food supply chain. Its basic function is containing food, facilitating its transport, and preventing any physical damage. Moreover, food packaging should preserve food safety and quality from the production stage until consumption [3]. Hence, the packaging acts as a barrier to protect the food from many environmental factors including oxygen, light, moisture, dust, pests, volatiles, and both microbiological and chemical contamination. Furthermore, packaging contributes to establishing convenient storage conditions for the consumer, which reduces food degradation [4].

Food packaging's role is continuously being ameliorated in response to consumer needs. Who are always willing for healthier, safer, and higher quality foods with long shelf life. In this optic active packaging (AP) was developed as a novel method of food preservatives [5]. Taking into account that AP is defined by the European regulation (EC) No 450/2009, as systems designed to *deliberately incorporate components that would release or absorb substances into or from the packaged food or the environment surrounding the food* [6]. Active packaging including antioxidant packaging, antimicrobial packaging, moisture absorbers, carbon dioxide emitters, and ethanol emitters. AP systems can be subdivided into active releasing systems (emitters) and active scavenging system (absorbers) [5]. AP scavenging systems are oxygen scavenger, moisture scavenger, and ethylene absorber. The presence of moisture in packaging affects food quality including appearance and texture. Several desiccants such as Zeolits and silica are usually used to prevent these problems [7]. Ethylene is responsible for chlorophyll degradation; moreover, it may be incriminated in shortening life of leafy products. Therefore, using ethylene scavengers, including Zeolites, nanoparticles and potassium permanganate can prevent food degradation [7].

The presence of oxygen in packaging can lead to product oxidation and the development of several aerobic microorganisms, resulting in the loss of some nutritional elements and the modification of color and taste. That is why using oxygen scavenging (OS) agents is a useful solution. Oxygen-scavenging gents can essentially include metallic, organic, polymer-based, or enzyme-based [8].

Metallic OS such as iron powder, activated iron, ferrous oxide, Co (II), iron salt, and Zn are oxidized in the presence of moisture. Iron-based OS are the most used agents for the conservation of packaged food, due to their low price and efficiency. However, using iron-based scavenging films have some drawbacks such as metal contamination of food and the reduction of its efficiency in high temperature [8]. Hence, researchers are substituting iron-based OS by organic agents (OA) such as tocopherol, ascorbic acid, ascorbic acid salts, isoascorbic acid, catechol, hydroquinone, sorbose, lignin, polyunsaturated fatty acids gallic acid, characterized by their low-molecular-weight oligomers. These agents have the possibility to be added to oxygen-scavenging polymers or polymer films. Indeed, side chains react with oxygen, or the backbone is broken apart when the polymer reacts with oxygen. Despite the advantages of organic-based scavenging films, it presents several drawbacks such as its relatively high cost and lower scavenging activity [9]. Polymer-based oxygen scavengers presents a new side of OS agents such as polymer-metal complexes, polyolefin, and oxidation-reduction resins. The main drawback of polymer-based oxygen scavengers is the possibility of by-product generation, such as organic acids, ketones, or aldehydes, during the reaction between oxygen and polyunsaturated molecules such as fatty acids affecting the color and the flavor of food products [10]. Another approach is, also used for oxygen scavenging in food packaging is the use of enzyme such as the mixture glucose oxidase and catalase. United enzyme oxygen scavenging systems are sensitive to variations in water activity, pH, salt content, and temperature.

Recently, natural products gained in popularity over synthetic products because of their 'safe status' and their lower perceived risk. Several studies have shown that natural product sources including flavonoids, polyphenols, tocopherols, and essential oils (EOs), extracted from plants may be valuable in food industry. These molecules are reported to be potential candidates for being included in AP [11, 12]. EOs are characterized by their complex and rich composition, conferring them a potential antioxidant and antimicrobial activity [12, 13]. The present study will focus on the possibilities of using EOs as effective alternatives or complements to synthetic chemical compounds in the AP system.

2. Essential oils

Essential oils are defined by ISO as '*Product obtained from vegetable raw material either by distillation with water or steam or from the epicarp of Citrus fruits by a mechanical process, or-by dry distillation*' [14]. Buchbauer and his collaborators [15] also define them as '*more or less volatile substances with more or less odorous impact, produced either by steam distillation or dry distillation or by means of a mechanical treatment from one single species*'. Essential oils are famous for their rich composition, mainly aromatic and volatile compounds. Indeed, EOs are composed of 85–99% volatile and 1–15% nonvolatile compounds. Volatile compounds are mainly terpenoids, aldehydes, ketones, esters, methoxy derivatives, acids, alcohols, phenols (eugenol), and methylenedioxy compounds (myristicin). These compounds confer to EOs many biological activities and make them an integral part of everyday life [15, 16]. EOs composition depends on many factors including plant species age, genetic factors, time of harvest, season, and extraction method [11, 17]. EOs are utilized for perfume and cosmetic product formulation. Moreover, they are used in the formulation of deodorizers, air fresheners as well as in the formulation of several pharmaceutical products. In agriculture, EOs are used as biocides. Furthermore, EOs are mainly used in the food industry for their flavors and their antioxidant and antimicrobial properties [18]. Synthetic preservatives are used in the food industry but can induce allergic reactions, intoxications, and even cancer. Therefore, industries are looking for other alternatives such as plant extracts and EOs characterized by their potential antimicrobial and antioxidant activities. Recently, there is a growing interest in the use of EOs in food packaging and in food due to their Generally Regarded as Safe (GRAS) and their significant antioxidant and antimicrobial activities [18].

3. Essential oil incorporation technology in active packaging

The EOs, which are known for their biological properties, are increasingly being employed as natural preservatives in food packaging. This approach tends to limit the usage of synthetic additives, increasing consumer acceptance of safe products [19]. Indeed, the EOs compounds can be progressively released at a suitable rate from the active packaging into the atmosphere surrounding the food product, exerting their positive antibacterial and antioxidant effects and, therefore, increasing the shelf life of the food product [20].

However, the EOs are challenging to efficiently include in active food packaging due to their volatilization, insolubilization in water, and chemical instability. As a result, these active agents must be integrated into matrices to sustain their biological characteristics during packaging manufacturing and then during various stages of food transportation and storage [18, 21]. Furthermore, to enhance the EOs biological effects in active packaging, optimal retention, and sustained, controlled release are required. This latter feature is particularly essential in extending the shelf life of the foods by prolonging the release period of EO compounds for keeping a continuous biological activity [22].

Several approaches have been taken to develop controlled-release active packaging. All of them are based on the use of biodegradable polymers or copolymers with filmogenic properties, such as polysaccharides (cellulose, starch, and chitin), proteins (gelatin, zein, gluten, and casein), and lipids [23]. To these promising vehicle polymers, plasticizers, crosslinking, and reinforcing agents can be added to enhance the mechanical properties of film packaging [24].

The EOs can be included in polymer matrices by simply blending film ingredients and casting methods to film formation or by employing several encapsulation technologies for EOs incorporation in active packaging film matrices. The casting method is widely used in the production of film packaging [25–30]. It simply entails spreading over a flat surface a prepared film-forming solution containing an active ingredient such as EOs and a filomgenic polymer, both dissolved in a solvent. The solvent is then removed by drying. A plasticizer that changes three-dimensional organization, lowers attractive intermolecular interactions, and increases free volume and chain mobility is typically added to the basic film recipe. Glycerol is the most common plasticizer used for its stability and compatibility with hydrophilic biopolymers. As a result, the film has greater extensibility and flexibility, both of which are important in film design [31]. However, the direct incorporation of EO in active films via the blending and casting methods has several limitations. It has been stated that microencapsulated oregano EO in soy protein concentrate films provides emulsion-based products with better mechanical properties as well as antibacterial action against food pathogens compared to films containing free EO [32]. Therefore, EOs encapsulation, applied to active films, constitutes an interesting alternative for preserving the active agents. It consists of forming a physical barrier between the active agent and the surrounding environment, providing the created capsules physical and chemical stability as well as enhanced biological (antibacterial and antioxidant) and functional qualities (better handling) [33].

Encapsulation is the process of entrapping active agents (core materials as EOs) by another substance that serves as the wall material, resulting in nanometer, micrometer, or millimiter capsules [18, 34]. Furthermore, encapsulating EOs before forming active films is more efficient because it increases EO stability and bioavailability and enables controlled release to the external medium around the encapsulated particles by the diffusion process. This is the primary role of active film packaging in preserving food products [35]. Zhang and his collaborators [22] emphasized the necessity of gradual release of EO components and a prolonged sustained release time of EO to ensure the efficiency of antibacterial activity throughout the shelf life of food goods.

Microcapsules or nanocapsules can be generated depending on the encapsulation technology specificity. Several investigations on active agent nanoencapsulation have recently been published [33, 36, 37]. Kapustova and his collaborators [38] stated that the nano-range (10^{-9}) of nanocapsules, which are a thousand times smaller than microcapsules, increases the surface-to-volume area for better efficiency in the delivery of EOs to targeted locations with greater stability and dispersibility.

Encapsulating technologies were classified into two categories: those that use chemical processes such as complex coacervation, liposomes, solid nanoparticles, and ionic gelation and those that rely on physical processes such as spray drying, extrusion, and solvent removal [39]. In general, more than one of the above technologies is often used to produce the desired microcapsules [40].

Furthermore, emulsifying the EOs compounds is usually required prior to encapsulation; it is considered as a preencapsulation step [41]. To stabilize the emulsion, high shear [42], or high pressure [43] or ultrasonication [44] must be used to homogenize the wall-core material. Nevertheless, droplets (oil in water) have such a loose structure they cannot effectively protect active substances; therefore, they must be immobilized in a solid matrix [45].

3.1 Encapsulation methods based on physical processes

Spray drying is commonly used to encapsulate EOs compounds [43, 46, 47]. It is primarily based on a three-step process: (1) wall-core material dispersion preparation, (2) wall-core material dispersion homogenization, and (3) dispersion atomization and drying [18]. The wall material must be carefully selected to improve EO component retention while also preventing oxidative alterations and volatilization [18]. Whey protein is commonly used as a wall material to facilitate emulsion formation and interfacial stabilization. Other additives, such as maltodextrins, can be employed to aid the encapsulation process, resulting in a bigger crust surrounding the drops and adequate oxidation protection [48].

Talon and his collaborators [43] examined whey protein and lecithin as wall materials for spray drying encapsulation of eugenol (7%). Both have been found to be effective against antioxidant and antibacterial properties when tested on *Escherichia coli* and *Listeria innocua*. Zhang and his collaborators [47] used spray drying technology to encapsulate a mixture of three EOs in gelatin-chitosan: cinnamon (*Cinnamomum cassia*), peppermint (*Mentha haplocalyx*), and lemon (*Citrus limon*). The EO combination exhibited a synergistic antibacterial activity based on the cooperation of different EOs.

Extrusion is often used to encapsulate active agents [34]. It works by forcing a substance through an orifice of varying width and shape according on the desired capsules [49]. Five extrusion technologies are used based on extruder specificity and other parameters: (1) hot-melt extrusion, (2) melt injection, (3) centrifugal/co-extrusion, (4) electrostatic/electrospinning, and (5) particle from gas-saturated solution. In hot-melt extrusion, the wall material is first introduced to the extruder, after plasticized, the active agent is applied to promote the interaction of the wall and core materials. In melt-injection, the active agent is directly dispersed in the heated wall material (80–140°C), then pressed through orifices into a bath of cold solvent to allow capsule solidification. Both extrusion processes required polymers with high flow characteristics and active agents that could endure high temperatures.

In co-extrusion technology, the wall and core materials are introduced separately through several nozzles located on the extruder's exterior surface. The wall material and active agent come into contact at the interface due to centrifugal forces, resulting in a polymerization reaction and the formation of microcapsules [34]. Because it requires less energy for encapsulation, this co-extrusion method is particularly suited for EOs compounds and probiotic bacteria as active agents.

Electrostatic extrusion, known as electrospinning, is a one-step process for producing micro and nanocapsules [50]. The introduction of an electric field between a charged needle (containing the microcapsules) and the collecting solution causes the microcapsules to be unable to stand at the mouth of the needle, resulting in the formation of a charge stream of small drop [34]. Since, electrospinning operates at ambient temperature and atmospheric strain, it is particularly suited to the encapsulation of EOs compounds [51].

3.2 Encapsulation methods based on chemical processes

Complex coacervation is largely employed for active agent encapsulation such as EOs [44, 52, 53]. It is mostly achieved by electrostatic forces of attraction between at least two polymers with opposite charges in aqueous fluids, with small contributions from hydrogen bonding, van der Waals forces, and hydrophobic interactions. As a result, the colloidal system separates into two liquid phases: one polymer-enriched precipitate phase and one polymer-depleted precipitate phase [54].

Polymers involved in coacervation are proteins and polysaccharides as wall material for the encapsulation of the active agent. This one is incorporated through emulsification in wall material to provide stability and protection. Finally, the capsules are separated using a physicochemical environment destabilization (pH and temperature) [18].

Ban and his collaborators [44] used coacervation to encapsulate ginger EOs in a mixture of chitosan (CH) and sodium carboxymethyl cellulose (CMC). The microcapsules with the same CH/CMC ratio have a crosslinking structure that may bind EOs, resulting in a high encapsulation efficiency (88.5%) and retention rate of volatile EO release to extend jujube shelf life.

Cyclodextrins (CDs) have been regarded as one of the preferred encapsulating polymers in the pharmaceutical industry and, recently, in the food industry [55]. CDs are a distinct family of molecules that are produced naturally through the degradation of starchy compounds. They are classified into three types: α-, β-, and γ-cyclodextrins and made up of D-glucose units linked together by glycocidic bonds between α-(1,4) carbon atoms. The toroidal structure of these compounds provides a hydrophobic interior cylindrical cavity and hydrophilic sides. As a result, the central cavity can form a stable combination with a guest molecule [37, 56]. EOs are well suited to being encapsulated in cyclodextrins, and so remaining protected while being released from the inclusion complex at a controlled rate, which is ideal for active packaging applications.

Silva and his collaborators [55] recently developed CD polymers such as CD nanosponges (CD-NS). These nanosponges are innovative crosslinked cyclodextrin polymers nanostructured within a three-dimensional network. They provide better stability and formulation flexibility with sustained release.

The technology of ionic gelation encapsulation has received a lot of attention in recent years because of its high adaptability to many types of active agents and low-cost approach [32]. This technology is particularly useful for encapsulating EO compounds to protect them from environmental deterioration [57, 58]. It is based on the ionic crosslinking of a polymer solution containing the active substance to encapsulate in the presence of multivalent cations [59]. As a result, complexation between oppositely charged species occurs under continual agitation [39]. Alginate and chitosan are the two most common coating materials. Both are nontoxic, highly biocompatible polymers with good mechanical resistance, making them appropriate for active food packaging applications [23].

Solid lipid nanoparticles (SLNs) are gaining popularity as attractive carriers for bioactive agents, particularly those with a lipophilic character such as EOs. SLNs are nanometer-sized colloidal particles formed from oil-in-water emulsions containing lipids that are solidified at room temperature and stabilized by the addition of a surfactant. The advantages of SLNs over other encapsulation methods include their biodegradability, gradual degradation, sustained release of active agent, and

EO incorporation technology in AP	Essential oil	Active packaging matrices	References
Blending Casting method	Torch ginger (*Etlingera elatior Jack*) inflorescence	Torch ginger EO (0.1–0.8%) was incorporated into starch solution at 3% (w/v) and 0.3% of glycerol (w/v) as plasticizer The active film was tested on chicken meat	[29]
Encapsulation • Emulsion Casting method	Apricot (*Prunus armeniaca*) kernel	Apricot kernet EO was incorporated to chitosane with acetic acid as a solvent and Tween 80 The film was tested for bread slices	[26]
	Lemongrass (*Cymbopogon citratus* L.)	Lemongrass EO was incorporated into two different formulations of biopolymer emulsions chitosan-gelatin and pectin-gelatin. Glycerol and Tween 80 were added as a plasticizer and an emulsifier, respectively, to the biopolymer solution The film was tested on storage of raspberries	[61]
Encapsulation • Emulsion • Ion gelation Casting method	Lemon	Emulsion was produced with Chitosan, Tween, and lemon. Then, tripolyphosphate solution was added to form nanocapsules The freeze-dried nanocapsules were added to Grass carp collagen to prepare edible films The film was tested on storage of chilled pork	[24]
Encapsulation: Spray drying	Mixture of three EOs:Cinnamon (*Cinnamomum cassia*), peppermint (*Mentha haplocalyx*), and lemon (*Citrus limon*)	A mixture of three EOs cinnamon, peppermint, and lemon (*Citrus limon*) were incorporated in gelatin-chitosan material by spray drying	[47]
Encapsulation: • Extrusion: melt-injection	• *Rose (Rosa eglanteria) seed EO* • *Ginger (Zingiber officinalis) root*	Low-density polyethylene (LDPE) pellets and active agents were mixed and melted together, then cooled and cut into granules. The extruded granules were fed to a blown film extrusion machine The film was tested on fresh meat	[62]
Encapsulation • Emulsion • Coacervation • Freeze drying	Ginger	Ginger EO was incorporated in polysaccharides, a mixture of chitosan and sodium carboxylmethyl cellulose The microcapsules were tested on postharvest jujube fruit	[44]

EO incorporation technology in AP	Essential oil	Active packaging matrices	References
Encapsulation • β-Cyclodextrin • Electrospinning	Cinnamon	Cinnamon EO (0.5–3 g) was encapsulated in polyvinyl alcohol and β-cyclodextrin polymers, producing nanofibrous films The nanofibrous film was tested on the preservation of fresh strawberries at 4°C for 18 days	[50]
Encapsulation: • Electrospinning	*Thyme*	*Thyme EO was incorporated in a zein solution with 30% (w/v)* of zein in glacial acetic acid *Zein nanofiber film was tested for the packaged strawberries*	[63]

Table 1.
Technological trends of active film packaging containing EO.

greater encapsulation effectiveness for lipophilic substances [60]. **Table 1** reports technological trends of active films packaging containing EO in the last 6 years (ScienceDirect—Elsevier).

4. Incorporation of EOs in active packaging system as antimicrobial agent

Several microorganisms are a threat for food sustainability and human health. They are implicated in quantitative and qualitative food loss [64]. Indeed, in 2011, it was estimated by the Food and Drug Administration (FDA) that 1.3 billion tons of food are annually discarded owing to their contamination by microbial spoilage, contributing to food insecurity and financial losses. Despite the evolution of production and packaging techniques, food contamination by pathogenic microorganisms has remained a persistent global challenge. Currently, there is a growing interest in integrating EOs as antimicrobial agents in food and food packaging [18]. Many previous studies have found that several EOs are efficient against diverse microbes including yeast, bacteria, fungi, and viruses [16, 18, 65].

Some EOs compounds such as aldehydes, phenols and oxygen-containing terpenes especially compound-containing phenol groups are characterized by their potential antibacterial activity [66, 67]. Indeed, phenolic compounds are implicated in the destruction of the bacterial cell membrane and permeability. The hydroxyl groups carried by the phenolic compound are implicated in the inhibition of microorganism enzyme activity. Eugenol and carvacrol are also reported to be potential antimicrobial agents [20].

Previous works showed that antibacterial activity of EOs depends on many factors including the bacterial cell wall composition [68] and cellular shape [65]. Indeed, it was shown that Gram-negative bacteria are more resistant than Gram-positive bacteria to EOs [68]; moreover, rod-shaped bacteria are more susceptible to EOs than cocci [69]. The antibacterial activity of EOs is associated with their lipophilic nature, enabling their accumulation in membranes, and making the membrane their main target [68].

To avoid food deterioration by microbial contamination, antimicrobial agent can be mixed into the initial food formulations, which may affect the taste of the

Essential oil	Microorganisms	References
Cinnamon and clove EOs	*Penicillium commune and Eurotium amstelodami*	[72]
Clove buds	*Staphylococcus aureus*, *Listeria monocytogenes*, *Salmonella typhimurium*, and *Escherichia coli*	[73]
Oregano and garlic EOs	*Escherichia coli*, *Salmonella enteritidis*, *Listeria monocytogenes*, *Staphylococcus aureus*, and *Penicillium* spp	[74]
Trachyspermum ammi EO (Ajowan)	*Pseudomonas* spp., *Staphylococcus aureus*	[75]
Cinnamon, melaleuca and citronella EOs	*Salmonella* sp., *Pseudomonas aeruginosa*, *Staphylococcus aureus*, *Escherichia coli*, *Aspergillus niger*, and *Staphylococcus epidermidis*	[76]
Cinnamon EO	*Listeria monocytogenese*, *Listeria garayi*	[77]
Thyme EO	*Escherichia coli*	[47]
Torch ginger (*Etlingera elatior* Jack) inflorescence EO	*Bacillus subtilis*, *Staphylococcus aureus*, *Listeria monocytogenes*, *Salmonella typhimirium*, and *Escherichia coli*	[29]
Laurus nobilis L. leaf EO)	*Bacillus cereus* and *Salmonella typhimurium*	[78]
Anise (*Pimpinella anisum* L.)	*Escherichia coli*, *Staphylococcus aureus*, *Saccharomyces cerevisiae*, and *Aspergillus niger*	[79]
Mixture of three EOs: Cinnamon (*Cinnamomum cassia*), Peppermint (*Mentha haplocalyx*), and lemon (*Citrus limon*)	*E. coli*, *Salmonella typhimurium* and *Staphylococcus aureus*	[21]

Table 2.
Incorporation of EOs in active packaging system as antimicrobial agent.

food [70]. Antimicrobial agents, including EOs can be also applied directly by brushing, dipping or spraying on the food surface. Nevertheless, active agent of antimicrobial substances may be evaporated, inactivated or can migrate into the bulk of the foods [71]. Thus, the incorporation of t EO on the packaging film, where they are gradually released to the food surface reducing the contamination and the development of microbial agent, provide better efficiency for foodstuffs preservation. **Table 2** illustrates many recent works published in the last ten years using EOs as antimicrobial agents in active packaging.

5. Incorporation of EOs in active packaging system as antioxidant agent

Food oxidative damage is initiated by the interactions of reactive oxygen species (ROS) including superoxide radicals (O_2^{2-}), hydrogen peroxide (H_2O_2), and hydroxyl radicals (OH^-) with oxidizable compounds. Food oxidative damages usually are implicated in shortening food shelf life, loss of color, odor, and flavor and lowering nutritional value [80].

Amorati and his collaborators [81] have defined an antioxidant compound by its ability to slow or retarding the oxidation of another material allowing protection from oxidative stress. Antioxidants can be classified into two main groups (preventive antioxidants and chain-breaking antioxidants) depending on their mechanism of action. Preventive antioxidants inhibit the initiation of radical species formation processes such as enzymes (catalase and superoxide dismutase) and metal chelators (phytic acid) [81]. Chain-breaking antioxidants inhibit or block autoxidation by reacting speeder than oxidizable substrate, forming neutral chemical species that cannot propagate the oxidation chain [81].

Thermoplastic films are used in packaging to exclude oxygen, avoiding the interaction between ROS and foods [82]. Nevertheless, using Thermoplastic films generates nondegradable packaging waste, and many new laws were proposed to reduce or ban single-use plastics [83].

Many previous studies have reported that EOs have a potential antioxidant activity contributing in the attenuation of free-radical oxidative reactions [13, 84]. EOs can prevent lipid oxidation in food through the inhibition of the food oxidative initiation, terminating peroxides, blocking the formation of singlet oxygen [85, 86].

Even though the large chemical diversity of EOs composition, the main components of common EOs can be classified into two structural families: terpenoids (monoterpene, sesquiterpene, diterpene) and phenylpropanoid [81]. Both terpenoid and phenylpropanoid contain phenolic compounds, which are antioxidants owing to their high reactivity with peroxyl radicals [87]. Moreover, phenolic compounds, ethers, aldehydes, ketones, and certain alcohols can enhance the antioxidant properties of the EOs [88]. Eugenol and carvacrol are also reported to be potential antioxidant agents [20].

Essential oil	Antioxidant effect of EO	References
Citrus sinensis (L.) Osbeck	Shelf life extension of Pink shrimp (*Parapenaeus longirostris*) by about n10 days	[92]
Peppermint EO	Up to the 45th day of storage at 40°C, Peppermint EO decreased considerably the formation of hydroperoxides in soybean oil at 40°C	[93]
Thyme EO	Amelioration of antioxidant activity and extension of the shelf life of chilled meat	[75]
Thyme EO	Increased antioxidant activity of strawberry fruits during the first 6 days of storage	[63]
Lemongrass EO	Active packaging film developed by the combination of lemongrass EO and chitosan protected the chicken patties packed from lipid peroxidation	[94]
Rosemary EO	Chitosan/sodium caseinate blend with 1% and 2% of rosemary EO reduced by 50% the malondialdehyde concentration of chicken meat	[42]
Ginger and grape seed EOs	• Meat shelf life extension by 6% and 2%, respectively, for ginger EO and grape seed EO • A positive effect on the freshness of meat was reported	[62]
Ginger EO	The antioxidant activities of the films significantly increased with the addition of ginger EO	[29]
Pepper-rosmarin	The addition of Pepper-rosmarin EO and poly (butylene adipate co-terephthalate) to active packaging film inhibited the oxidation of olive oil	[95]
Cinnamon EO	Increased antioxidant activity of silicon dioxide nanoparticles used in the active packaging	[96]

Table 3.
Incorporation of EOs in active packaging system as antioxidant agent.

EOs are mixtures of many compounds including different types of antioxidants or oxidizable components such as terpenoids and phenylpropanoid often coexist. Many previous works tried to study if the overall antioxidant activity of a natural EO can be attributed to the most effective antioxidant component. This hypothesis is true in some cases. However, many exceptions were found, reporting that EOs antioxidant activity is the result of the complex interaction among the oxidizable material to be protected and components. Generally, synergistic or antagonistic behavior is expected, depending on the composition of EOs and experimental conditions [89]. To prolong shelf life and prevent autoxidation of edible product, EOs characterized by their potential antioxidant activity can be used as a food ingredient, either as a part of active packaging. According to (EC) No 1333/2008 [90, 91]. EOs are considered as food additives when they represent a "substance not normally consumed as a food in itself and not normally used as a characteristic ingredient of food, whether or not it has nutritive value, the intentional addition of which to food for a technological purpose in the manufacture, processing, preparation, treatment, packaging, transport or storage of such food results, or may be reasonably expected to result, in it or its by-products becoming directly or indirectly a component of such foods". Moreover, EOs can also be considered as flavoring substances, according to (EC) No 1334/2008 [92] when they are 'products not intended to be consumed as such, which are added to food in order to impart or modify odor and/or taste; or products made or consisting of the following categories: flavoring substances, flavoring preparations, thermal process flavorings, smoke flavorings, flavor precursors or other flavorings or mixtures there of'. Many recent researches reported the efficiency of EOs used in active packaging as an antioxidant are reported in **Table 3**.

6. Conclusion

Food packaging protects food from environmental effect, contributing to the establishment of convenient storage conditions and reducing food degradation. In response to consumer demand for healthier and higher quality foods with long shelf life, food packaging is continuously being ameliorated. For this purpose, AP was developed. AP can be essentially classified into two categories: active releasing systems (emitters) and active scavenging systems (absorbers). In this study, we were interested in active releasing systems—specially EOs. Many approaches were used to develop controlled-release active packaging. Most of them are based on the use of biodegradable polymers or copolymers with filmogenic properties. EOs, can be incorporated into polymer matrices by blending, method, micro-nano encapsulation, or adsorption technologies. Several recent studies evaluated the effect of EO incorporation in AP as antibacterial or antioxidant agent, and it was found that most of the tested EO are implicated in extending food shelf life.

According to the most recent literature EOs incorporation in AP could be considered as an optimal alternative option in the food packaging industry to obtain healthier food with longer shelf life. However, additional studies for biological and organoleptic proprieties are required.

Conflict of interest

The authors declare that they have no conflicts of interest.

Author details

Imen Dridi[1*], Ahmed Landoulsi[1] and Nadia Smirani[2]

1 Faculty of Sciences of Bizerte, Laboratory of Biochemistry and Molecular Biology, University of Carthage, Bizert, Tunisia

2 Laboratory of Microbial Ecology and Technology LETMi-INSAT, The National Institute of Applied Science and Technology INSAT, Carthage University, Tunis, Tunisia

*Address all correspondence to: dridiimentun@gmail.com

References

[1] UNEP (United Nations Environment Programme). Food Waste Index Report 2021. Nairobi. 2021. pp. 99. ISBN No: 978- 92-807-3868-1. Available from: https://www.unep.org/resources/report/unep-food-waste-index-report-2021

[2] Pasarín V, Viinikainen T. Enabling a legal environment for the prevention and reduction of food loss and waste. In: Legal Brief. Rome, Italy: FAO; 2022. DOI: 10.4060/cc2278en

[3] Brennan L, Langley S, Verghese K, Lockrey S, Ryder M, Francis C, et al. The role of packaging in fighting food waste: A systematised review of consumer perceptions of packaging. Journal of Cleaner Production. 2021;**281**:125276. DOI: 10.1016/j.jclepro.2020.125276

[4] Bahrami A, Delshadi R, Assadpour E, Jafari SM, Williams L. Antimicrobial-loaded nanocarriers for food packaging applications. Advances in Colloid and Interface Science. 2020;**278**:102140. DOI: 10.1016/j.cis.2020.102140

[5] Yildirim S, Rocker B, Pettersen MK, Nygaard JN, Ayhan Z, Rutkaite R, et al. Active packaging applications for food. Comprehensive Reviews in Food Science and Food Safety. 2018;**17**(1):165-199. DOI: 10.1111/1541-4337.12322

[6] European Commission. Regulation (EC) No 450/2009 of the European Parliament and of the Council of 29 May 2009 on active and intelligent materials and articles intended to come into the contact with food. Official Journal of the European Union. 2009;**135**:3-11. Available from: https://eur-lex.europa.eu/LexUriServ/LexUriServ.do?uri=OJ:L:2009:135:0003:0011:EN:PDF

[7] Dey A, Neogi S. Oxygen scavengers for food packaging applications: A review. Trends in Food Sciences and Technology. 2019;**90**:26-34. DOI: 10.1016/j.tifs.2019.05.013

[8] Gaikwad KK, Singh S, Lee YS. Oxygen scavenging films in food packaging. Environmental Chemistry Letters. 2018;**16**:523-538. DOI: 10.1007/s10311-018-0705

[9] Byun Y, Bae HJ, Whiteside S. Active warm-water fish gelatin film containing oxygen scavenging system. Food Hydrocolloids. 2012;**27**:250-255. DOI: 10.1016/j.foodhyd.2011.06.010

[10] Li H, Tung KK, Paul DR, Freeman BD, Stewart ME, Jenkins JC. Characterization of oxygen scavenging films based on 1, 4-polybutadiene. Industrial & Engineering Chemistry Research. ;**2012**(51):7138-7145. DOI: 10.1021/ie201905j

[11] Dhifi W, Bellili S, Jazi S, Bahloul N, Mnif W. Essential oils' chemical characterization and investigation of some biological activities: A critical review. Medicine. 2016;**3**(4):E25. DOI: 10.3390/medicines3040025

[12] Qiana M, Liuabc D, Zhanga X, Zhongpingd Y, Ismaila BB, Yeabc X, et al. A review of active packaging in bakery products: Applications and future trends. Trends in Food Science & Technology. 2021;**114**:459-471. DOI: 10.1016/j.tifs.2021.06.009

[13] Dridi I, Haouel-Hamdi S, Cheraief I, Mediouni Ben Jemâa J, Landoulsi A, Chaouch R. Tunisian *Lavandula dentata* (L) flowering tops essential oil: Chemical composition, antimicrobial, antioxidant and insecticidal activities. Journal of Essential Oil Bearing Plants. 2021;**24**(3):632-647. DOI: 10.1080/0972060X.2021.1944326

[14] ISO/DIS 9235.2: 1997. Aromatic Natural Raw Materials—Vocabulary. Geneva: International Standard Organization. Available from: https://www.iso.org/standard/51017.html

[15] Buchbauer G, Jirovetz L, Jáger W, Plank C. Studies on the biological effects of fragrance compounds and essential oils upon inhalation. Journal of Pharmaceutical Sciences. 1993;**82**(6):660-664. DOI: 10.1002/jps.2600820623

[16] Ben Salha G, Herrera Díaz R, Lengliz O, Abderrabba M, Labidi J. Effect of the chemical composition of free-terpene hydrocarbons essential oils on antifungal activity. Molecules. 2019;**24**(19):3532. DOI: 10.3390/molecules24193532

[17] Sowndhararajan K, Deepa P, Kim M, Park SJ, Kim S. A review of the composition of the essential oils and biological activities of Angelica species. Scientia Pharmaceutica. 2017;**85**(3):33. DOI: 10.3390/scipharm85030033

[18] Mukurumbira AR, Shellie Keast R, Palombo EA, Jadhay SR. Encapsulation of essential oils and their application in antimicrobial active packaging. Food Control. 2022;**136**:108883. DOI: 10.1016/j.foodcont.2022.108883

[19] Manzoor A, Yousuf B, Pandith JA, Ahmad A. Plant-derived active substances incorporated as antioxidant, antibacterial or antifungal components in coatings/films for food packaging. Applications Food Bioscience. 2023;**53**:102717. DOI: 10.1016/j.fbio.2023.102717

[20] Lopez-GomezA,Navarro-MartínezA, Martínez-Hernandez GB. Effects of essential oils released from active packaging on the antioxidant system and quality of lemons during cold storage and commercialization. Scientia Horticulturae. 2023;**312**:111855. DOI10.1016/j.scienta.2023.111855

[21] Zhang J, Zhang J, Huang X, Shi J, Muhammad A, Zhai X, et al. Study on cinnamon essential oil release performance based on pH-triggered dynamic mechanism of active packaging for meat preservation. Food Chemistry. 2023;**400**:134030. DOI: 10.1016/j.foodchem.2022.134030

[22] Zhang W, Jiang J, Rhim JW, Cao J, Jiang W. Effective strategies of sustained release and retention enhancement of essential oils in active food packaging films/coatings. Food Chemistry. 2022;**367**:130671. DOI: 10.1016/j.foodchem.2021.130671

[23] Rout S, Tambe S, Kumar Deshmukh R, Mali S, Cruz J, Prakash Srivastav PA, et al. Recent trends in the application of essential oils: The next generation of food preservation and food packaging. Trends in Food Science and Technology. 2022;**129**:421-439. DOI: 10.1016/j.tifs.2022.10.012

[24] Jiang Y, Lan W, Sameen DU, Ahmed S, Qin W, Zhang Q, et al. Preparation and characterization of grass carp collagen-chitosan lemon essential oil composite films for application as food packaging. Trends in Food Science and Technology. 2020;**129**:421-439. DOI: 10.1016/j.ijbiomac.2020.05.202

[25] Danneenberg GDS, Funck GD, Cruxen CEDS, Marques JDL, da Silva WP, Fiorentini AM, et al. Essential oil from pink pepper as an antimicrobial component in cellulose acetate film: Potential for application as active packaging for sliced cheese. LWT-Food Science and Technology. 2017;**81**:314-318. DOI: 10.1016/j.lwt.2017.04.002

[26] Priyadarshi R, Sauraj KB, Deeba F, Kulshreshtha A, Negi YS. Chitosan films incorporated with apricot (*Prunus armeniaca*) kernel essential oil as active food packaging material. Food Hydrocolloids. 2018;**85**:158-166. DOI: 10.1016/j.foodhyd.2018.07.003

[27] Cheng J, Wang H, Xiao F, Xia L, Li L, Sameen DU, et al. Functional effectiveness of double essential oils@ yam starch/microcrystalline cellulose as active antibacterial packaging. Biological Macromolecules. 2021;**186**:873-885. DOI: 10.1016/j.ijbiomac.2021.07.094

[28] Vishnu Priya N, Vinitha UG, Meenakshi SM. Preparation of chithsan based antimicrobial active food packaging film incorporated with *Plectranthus amboinicus* essential oil. Biocatalysis and Agricultural Biotechnology. 2021;**34**:102021. DOI: 10.1016/j.bcab.2021.102021

[29] Marzlan AA, Muhialdin BJ, Abedin NHZ, Manshoor N, Ranjith FH, Anzian A, et al. Incorporating torch ginger (*Etlingera elatior* JACK) inflorescence essential oil into starch-based edible film towards sustainable active packaging for chicken meat. Industrial Crops and Products. 2022;**184**:115058. DOI: 10.1016/j.indcrop.2022.115058

[30] Araújo MNP, Grisi CVB, Duarte CR, Bastos de Almeida YM, Vinhas GM. Active packaging of corn starch with pectin extract and essential oil of *Turmeric Longa Linn*: Preparation, characterization and application in sliced bread. International Journal of Biological Macromolecules. 2023;**226**:1352-1359. DOI: 10.1016/j.ijbiomac.2022.11.248

[31] Espitia PJP, Soares DFF, Teófil RF, Coimbra JSR, Vitor DM, Batista RA, et al. Physical-mechanical and antimicrobial properties of nanocomposite films with pediocin and ZnO nanoparticles. Carbohydrate Polymers. 2013;**94**:199-208. DOI: 10.1016/j.carbpol.2013.01.003

[32] Dos Santos IP, Vitoria Galindob MAS, de Medeirosc J, Yamashitad F, Dutra Alvime I, Grossoa CRF, et al. Comparative study of the properties of soy protein concentrate films containing free and encapsulated oregano essential oil. Food Packaging and Shelf Life. 2019;**22**:100419. DOI: 10.1016/j.fpsl.2019.100419

[33] Zanetti M, Carniel TK, DalcaCnton F, dos Anjos RS, Gracher Riella H, de Araújo PHH, et al. Use of encapsulated natural compounds as antimicrobial additives in food packaging. A brief review. Trends in Food Science & Technology. 2018;**81**:51-60. DOI: 10.1016/j.tifs.2018.09.003

[34] Bamidele OP, Emmambux MN. Encapsulation of bioactive compounds by "extrusion" technologies. Critical Reviews in Food Science and Nutrition. 2021;**61**:3100-3118. DOI: 10.1080/10408398.2020.1793724

[35] Aloui H, Khwaldia K, Licciardello F, Mazzaglia A, Muratore G, Hamdi M, et al. Efficacy of the combined application of chitosan and locust bean gum with different citrus essential oils to control postharvest spoilage caused by *Aspergillus flavus* in dates. International Journal of Food Microbiology. 2014;**170**:21-28. DOI: 10.1016/j.ijfoodmicro.2013.10.017

[36] Khezerlou A, Jafari SM. Nanoencapsulated bioactive components for active food packaging. In: Handbook of Food Nanotechnology. 2020. pp. 493-532. DOI: 10.1016/B978-0-12-815866-1.00013-3

[37] Swain SS, Paidesetty SK, Padhy RN, Hussain T. Nano-technology platforms to

increase the antibacterial drug suitability of essential oils: A drug prospective assessment. OpenNano. 2023;**9**:100115. DOI: 10.1016/j.onano.2022.100115

[38] Kapustová M, Granata G, Napoli E, Puškárová A, Bucková M, Pangallo D, et al. Nanoencapsulated essential oils with enhanced antifungal activity for potential application on agri-food, material and environmental fields. Antibiotics. 2021;**10**:31. DOI: 10.3390/antibiotics10010031

[39] Reis DR, Ambrosi A, Luccioa MD. Encapsulated essential oils: A perspective in food preservation. Future Foods. 2022;**5**:100126. DOI: 10.1016/j.fufo.2022

[40] Zhao Y, Wang Y, Zhang Z, Li. Advances in controllable release essential oil microcapsules and their promising applications. Molecules. 2023;**28**(13):4979. DOI: 10.3390/molecules28134979

[41] Salvia-Trujillo L, Rojas-Graü A, Soliva-Fortuny R, Martín-Belloso O. Physicochemical characterization and antimicrobial activity of food-grade emulsions and nanoemulsions incorporating essential oils. Food Hydrocolloids. 2015;**43**:547-556. DOI: 10.1016/j.foodhyd.2014.07.012

[42] Fiore A, Park S, Volpe S, Torrieri E, Masi P. Active packaging based on PLA and chitosan-caseinate enriched rosemary essential oil coating for fresh minced chicken breast application. Food Packaging and Shelf Life. 2021;**29**:100708. DOI: 10.1016/j.fpsl.2021.100708

[43] Talón E, Lampi A-M, Vargas M, Chiralt A, Jouppila K, González-Martínez C. Encapsulation of eugenol by spray-drying using whey protein isolate or lecithin: Release kinetics, antioxidant and antimicrobial properties. Food Chemistry. 2019;**295**:588-598. DOI: 10.1016/j.foodchem.2019.05.115

[44] Ban Z, Zhang J, Li L, Luo Z, Wang Y, Yuan Q, et al. Ginger essential oil-based microencapsulation as an efficient delivery system for the improvement of Jujube (*Ziziphus jujuba Mill.*) fruit quality. Food Chemistry. 2020;**306**:125628. DOI: 10.1016/j.foodchem.2019.125628

[45] Bušić A, Komes D, Belšćak-Cvitanovic A, Cebin AV, Špoljarić I, Mršić G, et al. The potential of combined emulsification and spray drying techniques for encapsulation of polyphenols from rosemary (*Rosmarinus officinalis* L.) leaves. Food Technology and Biotechnology. 2018;**56**(4):494-505. DOI: 10.17113/ftb.56.04.18.5680

[46] Bao H, Ding HH, Charles APR, Hui D, Rakshit S, Nahashon S, et al. Application of yellow mustard mucilage in encapsulation of essential oils and polyphenols using spray drying. Food Hydrocolloids. 2023;**143**:108815. DOI: 10.1016/j.foodhyd.2023.108815

[47] Zhang Y, Yang S, Hardie WJ, Li X, Xiao M, Huang T, et al. Microcapsules of a cinnamon, peppermint, and lemon essential oil mix by spray: Preparation, characterization and antibacterial functions. Food Hydrocolloids. 2023;**145**:109103. DOI: 10.1016/j.foodhyd.2023.109103

[48] Sheu TY, Rosengerg M. Microstructure of microcapsules consisting of whey proteins and carbohydrates. Food Science. 1998;**63**:491-494 10.1111/j.1365-2621.1998.tb15770.x

[49] Alam MS, Kaur J, Khaira H, Gupta K. Extrusion and extruded products: Changes in quality attributes as affected by extrusion process parameters:

A review. Critical Reviews in Food Science and Nutrition. 2016;**56**(3):445-473. DOI: 10.1080/10408398.2013.779568

[50] Wen P, Zhu DH, Wu H, Zong MH, Jing YR, Han SY. Encapsulation of cinnamon essential oil in electrospun nanofibrous film for active food packaging. Food Control. 2016;**59**:366-376. DOI: 10.1016/j.foodcont.2015.06.005

[51] Vega-Lugo AC, Lim L-T. Controlled release of allyl isothiocyanate using soy protein and poly(lactic acid) electrospun fibers. Food Research International. 2009;**41**:933-940. DOI: 10.1016/j.foodres.2009.05.005

[52] Heckert Bastos LP, Vicente J, Corrêa dos Santos CH, Geraldo de Carvalho M, Garcia-Rojas EE. Encapsulation of black pepper (*Piper nigrum* L.) essential oil with gelatin and sodium alginate by complex coacervation. Food Hydrocolloids. 2019;**102**:105605. DOI: 10.1016/j.foodhyd.2019.105605

[53] Plati F, Paraskevopoulou A. Hemp protein isolate-gum Arabic complex coacervates as a means for oregano essential oil encapsulation. Comparison with whey protein isolate-gum Arabic system. Food Hydrocolloids. 2023;**136**:108284. DOI: 10.1016/j.foodhyd.2022.108284

[54] Muhoza B, Xia S, Zhang X. Gelatin and high methyl pectin coacervates crosslinked with tannic acid: The characterization, rheological properties, and application for peppermint oil microencapsulation. Food Hydrocolloids. 2019;**97**:105174. DOI: 10.1016/j.foodhyd.2019.105174

[55] Silva F, Caldera F, Trotta F, Nerín C, Domingues FC. Encapsulation of coriander essential oil in cyclodextrin nanosponges: A new strategy to promote its use in controlled-release active packaging. Innovative Food Science and Emerging Technology. 2019;**56**:102177. DOI: 10.1016/j.ifset.2019.102177

[56] Narayanan G, Shen J, Matai I, Sachde A, Boy R, Tonelli AE. Cyclodextrin-based nanostructures. Progress in Materials Science. 2022;**124**:100869. DOI: 10.1016/j.pmatsci.2021.100869

[57] Rajkumar V, Gunasekaran C, Paul A, Dharmaraj J. Development of encapsulated peppermint essential oil in chitosan nanoparticles: Characterization and biological efficacy against stored-grain pest control. Pestcide Biochemistry and Physiology. 2020;**170**:104679. DOI: 10.1016/j.pestbp.2020.104679

[58] Odjo K, Al-Maqtari QA, Yu H, Xie Y, Guo Y, Li M, et al. Preparation and characterization of chitosan-based antimicrobial filmscontaining encapsulated lemon essential oil by ionic gelationand cranberry juice. Food Chemistry. 2022;**397**(15):133781. DOI: 10.1016/j.foodchem.2022.133781

[59] Yeo Y, Park K. Control of encapsulation efficiency and initial burst in polymeric microparticle systems. Archives of Pharmacal Research. 2004;**27**:1-12. DOI: 10.1007/BF02980037

[60] Sharma M, Gupta N, Soni R, Gautam A, Reddy KR, Sharma S, et al. Recent advances in polymeric and solid lipid–based nanoparticles for controlled drug delivery. Nanomaterials in Diagnostic Tools and Devices. 2020;**2020**:131-158. DOI: 10.1016/B978-0-12-817923-9.00005-5

[61] Jovanović J, Krnjajić S, Ćirković J, Radojković A, Popović T, Branković G, et al. Effect of encapsulated lemongrass (*Cymbopogon citratus* L.) essential oil against potato tuber moth *Phthorimaea*

operculella. Crop Protection. 2020;**132**:105109. DOI: 10.1016/j.cropro.2020.105109

[62] Wrona M, Silva F, Salafranca J, Nerín C, Jose Alfonso M, Caballero MA. Design of new natural antioxidant active packaging: Screening flowsheet from pure essential oils and vegetable oils to ex vivo testing in meat samples. Food Control. 2021;**120**:107536. DOI: 10.1016/j.foodcont.2020.10753

[63] Ansarifar E, Moradinezhad F. Preservation of strawberry fruit quality via the use of active packaging with encapsulated thyme essential oil in zein nanofiber film. International Journal of Food Science and Technology. 2021;**56**:4239-4247. DOI: 10.1111/ijfs.15130

[64] Gonelimali FD, Lin J, Miao W, Xuan J, Charles F, Chen M, et al. Antimicrobial properties and mechanism of action of some plant extracts against food pathogens and spoilage microorganisms. Frontiers in Microbiology. 2018;**9**:1639. DOI: 10.3389/fmicb.2018.01639

[65] Wu J, Sun X, Guo X, Ge S, Zhang Q. Physicochemical properties, antimicrobial activity and oil release of fish gelatin films incorporated with cinnamon essential oil. Aquaculture and Fisheries. 2017;**2**(4):185-192. DOI: 10.1016/j.aaf.2017.06.004

[66] Ghabraie M, Dang VK, Tata L, Salmieri S, Lacroix M. Antimicrobial effect of essential oils in combinations against five bacteria and their effect on sensorial quality of ground meat. LWT—Food Science and Technology. 2016;**66**:332-339. DOI: 10.1016/j.lwt.2015.10.055

[67] Alsherbiny MA, Abd-Elsalam WH, El Badawy SA, Taher E, Fares M, Torres A, et al. Ameliorative and protective effects of ginger and its main constituents against natural, chemical and radiation-induced toxicities: A comprehensive review. Food and Chemical Toxicology. 2019;**123**:72-97. DOI: 10.1016/j.fct.2018.10.048

[68] Nazzaro F, Fratianni F, De Martino L, Coppola R, De Feo V. Effect of essential oils on pathogenic bacteria. Pharmaceuticals. 2013;**6**(12):1451-1474. DOI: 10.3390/ph6121451

[69] Lopez-Romero JC, Gonzalez-Ríos H, Borges A, Simoes M. Antibacterial effects and mode of action of selected essential oils components against *Escherichia coli* and *Staphylococcus aureus*. Evidence-based Complementary and Alternative Medicine. 2015;**2015**:795435. DOI: 10.1155/2015/795435

[70] Uz M, Altinkaya SA. Development of mono and multilayer antimicrobial food packaging materials for controlled release of potassium sorbate. LWT Food Science and Technology. 2011;**44**:2302-2309. DOI: 10.1016/j.lwt.2011.05.003

[71] Quintavalla S, Vicini L. Antimicrobial food packaging in meat industry. Meat Science. 2002;**62**:373-380. DOI: 10.1016/s0309-1740(02)00121-3

[72] Souza AC, Goto GEO, Mainardi JA, Coelho ACV, Tadini CC. Cassava starch composite films incorporated with cinnamon essential oil: Antimicrobial activity, microstructure, mechanical and barrier properties. LWT—Food Science and Technology. 2013;**54**:346-352. DOI: 10.1016/j.lwt.2013.06.017

[73] Hadidia M, Pouraminb S, Adinepourc F, Haghanib S, Jafaric SM. Chitosan nanoparticles loaded with clove essential oil: Characterization, antioxidant and antibacterial activities. Carbohydrate Polymers. 2020;**236**:116075. DOI: 10.1016/j.carbpol.2020.116075

[74] Seydim AC, Sarikus-Tutal G, Sogut E. Effect of whey protein edible films containing plant essential oils on microbial inactivation of sliced Kasar cheese. Food Packaging and Shelf Life. 2020;**26**:100567. DOI: 10.1016/j.fpsl.2020.100567

[75] Zhang Y, Zhou L, Zhang C, Show PL, Du A, Fu JC, et al. Preparation and characterization of curdlan/polyvinyl alcohol/thyme essential oil blending film and its application to chilled meat preservation. Carbohydrate Polymers. 2020;**247**:116670. DOI: 10.1016/j.carbpol.2020.116670

[76] Rech CR, Brabes KCS, Silva BEB, Martines MAU, Silveira TFS, Alberton J, et al. Antimicrobial and physical–mechanical properties of polyhydroxybutyrate edible films containing essential oil mixtures. Journal of Polymers and the Environment. 2021;**29**:1202-1211. DOI: 10.1007/s10924-020-01943-0

[77] Tao R, Sedman J, Ismail A. Antimicrobial activity of various essential oils and their application in active packaging of frozen vegetable products. Food Chemistry. 2021;**360**:129956. DOI: 10.1016/j.foodchem.2021.129956

[78] Kurtfaki M, Yalcin MY. Characterization of *Laurus nobilis* L. leaf essential oil incorporated maize starch and rice protein films. Journal of Food Measurement and Characterization. 2023;**78**:4954-4962. DOI: 10.1007/s11694-023-02013-4

[79] Jafari R, Zandi M, Ganjloo A. Characterization of alginate-Gelatin edible film containing Anise (*Pimpinella anisum* L.) essential oil. Journal of Polymers and the Environment. 2023;**31**:1568-1583. DOI: 10.1007/s10924-022-02707-8

[80] Shahidi F, Ambigaipalan P. Phenolics and polyphenolics in foods, beverages and spices: Antioxidant activity and health effects. Journal of Functional Foods. 2015;**18**:820-897. DOI: 10.1016/j.jff.2015.06.018

[81] Amorati R, Foti MC, Valgimigli L. Antioxidant activity of essential oils. Journal of Agricultural Food Chemistry. 2013;**61**:10835-10847. DOI: 10.1021/jf403496k

[82] Atarés L, Chiralt A. Essential oils as additives in biodegradable films and coatings for active food packaging. Trends in Food Science and Technology. 2016;**48**:51-62. DOI: 10.1016/j.tifs.2015.12.001

[83] European Commission. Proposal for a Directive of the European Parliament and of the Council on the Reduction of the Impact of Certain Plastic Products on the Environment. Brussels, Belgium: European Commission; 2018. Available from: https://eur-lex.europa.eu/resource.html?uri=cellar:4d0542a2-6256-11e8-ab9c-01aa75ed71a1.0001.02/DOC_1&format=PDF

[84] Carpena M, Nuñez-Estevez B, Lopez AS, Oliveira PG, Prieto MA. Essential oils and their application on active packaging systems. Resources. 2021;**10**:7. DOI: 10.3390/resources10010007

[85] Manessis G, Kalogianni AI, Lazou T, Moschovas M, Bossis I, Gelasakis AI. Plant derived natural antioxidants in meat and meat products. Antioxidants. 2020;**9**(12):1215. DOI: 10.3390/antiox9121215

[86] Shu Q, Niu YW, Zhao WJ, Chen QH. Antibacterial activity and mannosylerythritol lipids against vegetative cells and spores of *Bacillus cereus*. Food Control. 2019;**106**:106711. DOI: 10.1016/j.foodcont.2019.106711

[87] Foti M. Antioxidant properties of phenols. Journal of Pharmacy and Pharmacology. 2007;**59**:1673-1685. DOI: 10.1211/jpp.59.12.0010

[88] Amjadi S, Almasi H, Ghadertaj A, Mehryar L. Whey protein isolate-based films incorporated with nanoemulsions of orange peel (*Citrus sinensis*) essential oil: Preparation and characterization. Journal of Food Processing and Preservation. 2021;**45**(2):e15196. DOI: 10.1111/jfpp.15196

[89] Kulisic T, Radonic A, Milos M. Inhibition of lard oxidation by fractions of different essential oils. Grasas y Aceites. 2005;**56**:284-291. DOI: 10.3989/gya.2005.v56.i4.94

[90] European Parliament and the council of 16 December 2008 Union. Regulation (EC) No 1333/2008 on food additives. Official Journal of the European Union. 2008;**34**:67-84

[91] European Commission. Regulation (EC) No 1334/2008 of the European Parliament and of the Council of 16 December 2008 on flavourings and certain food ingredients with flavouring properties for use in and on foods and amending Council Regulation. (EC) No 110/2008 and Directive 2000/13/EC. Official Journal of the European Union. 2008;**354**:34-49

[92] Alparslan Y, Yapici HH, Metin C, Baygar T, Günlü A, Baygar T. Quality assessment of shrimps preserved with orange leaf essential oil incorporated gelatin. LWT- Food Science and Technology. 2016;**72**:457-466. DOI: 10.1016/j.lwt.2016.04.066

[93] Dastgerdi GF, Goli SAH, Kadivar M. A new antioxidant active film based on HDPE and peppermint essential oil for packaging soybean oil. Journal of the American Oil Chemists Society. 2016;**93**:657-664. DOI: 10.1007/s11746-016-2806-9

[94] Contini LRF, Zerlotini TS, Brazoli IF, dos Santos JWS, Silva MF, Lopes PS, et al. Antioxidant chitosan film containing lemongrass essential oil as active packaging for chicken patties. Journal of Food Processing and Preservation. 2021;**46**:e16136. DOI: 10.1111/jfpp.16136

[95] De Andrade MF, Silva MG, Silva IDL, Caetano VF, Filho LEPTM, Vinhas GM, et al. Pepper-rosmarin essential oil (*Lippia sidoides* Cham.) as an antioxidant additive for PBAT – poly (butylene adipate coterephthalate) films and its application for active packaging. International Journal of Food Science and Technology. 2022;**57**:5966-5972. DOI: 10.1111/ijfs.15931

[96] Aguilar VM, Juárez MGP, Sanchez PCC, Zamora LL, Alvarado ED, Castañeda EG, et al. Evaluation of the antioxidant and antimicrobial potential of SiO2 modified with cinnamon essential oil (*Cinnamomum Verum*) for its use as a nanofiller in active packaging PLA films. Antioxidants. 2023;**12**:1090. DOI: 10.3390/antiox12051090

Chapter 10

Essential Oils and the Circular Bioeconomy

Elena Stashenko and Jairo René Martínez

Abstract

The average annual trade of over 250 thousand tons of essential oils generates over 250 million tons of distillation residues, posing environmental problems due to incineration or landfill overburden. The circular economy focuses on reducing resource inputs, waste generation, and pollution, for sustainability. Implementing circular economy principles not only mitigates environmental concerns but also creates economic opportunities by utilizing residual biomass. Nonvolatile secondary metabolites, like flavonoids and phenolic compounds, remain in plant material during essential oil distillation. These bioactive substances can be extracted from the biomass distillation residues. Instead of discarding or burning waste from essential oil production, it can be processed to make extracts. The residue can be converted into biochar, a carbon-rich material beneficial for soil improvement. Other end uses include generating combustible bio-oil and using distillation residues for mushroom cultivation. Circular economy practices in the essential oil agroindustry have implications beyond the field itself. By providing raw materials for various sectors and industries, such as agriculture, cosmetics, and pharmaceuticals, this agroindustry can contribute to broader sustainability goals. While the adoption of circular economy principles presents technological challenges, the potential benefits in terms of waste reduction, value addition, and sustainability justify ongoing research and development efforts.

Keywords: circular economy, essential oil, hydrosol, residual biomass, compost, flavonoid, polyphenol

1. Introduction

An annual average of 264 thousand tons of essential oils (HS code 3301) were exported worldwide during the period 2017–2021 [1]. The actual essential oil production figures should be higher because this annual average does not include the amounts not exported and consumed within the essential oil-producing countries. Since the essential oils are isolated from plant material with typical yields in the range of 0.5%, the yearly byproduct of the traded amounts of essential oils should have been approximately 52 million tons of vegetal material. The disposal of these large amounts of material leads to serious environmental problems associated with river pollution, incineration, or landfill overburden. Alternatively, this residual biomass may become a valuable source of products with reduced environmental impact if a highly recommended approach, the application of the principles of circular economy, is used to add value, generate energy, and reduce waste to zero. The technological challenges are massive, but the potential benefits of following the latter principles have been the subject of many scientific articles during the past decade, not only around essential oil production but also in many different trade sectors. Waste and byproducts from food and agro-industries may no longer be associated with pollution and could be transformed and used to fight hunger and malnutrition. This chapter presents examples of the application of circular economy principles at laboratory and production scales in the essential oil agroindustry.

2. Circular economy

The circular economy approach is a closed-loop system that aims to reduce the use of resource inputs, waste generation, pollution, and carbon emissions. It is a vision toward a sustainable society with great responsibility in the production of biomass and end-of-life product recovery. Reduction, repair, remanufacturing, and recycling are normally acknowledged as the representative loops of the circular economy [2]. Some authors refer to these as holistic approaches to add value to biowastes from fruit and vegetable processing [3], of which aromatic plants represent a small percentage. The goal of adding value and reducing waste to zero may invoke participation not only in physical processes related to extraction but also in chemical transformations (hydrolysis), fermentations, and bioprocessing with microorganisms. Some traditional or conventional extraction methods are not recommended due to time consumption, intensive labor demand, their use of large amounts of organic solvents, or elevated temperatures that degrade some compounds of interest. Alternative emerging technologies are faster and have reduced environmental impact. They include processes such as supercritical carbon dioxide extraction, subcritical water extraction, ultrasound mixing, microwave heating, electric pulse discharge, and enzymatic hydrolysis, which frequently lead to improved yields [4]. However, there is still a large room for improvement in scaling up their operation and changing from batch to continuous operation [5].

Plant secondary metabolites do not participate in a direct manner in basic functions such as growth and development but are very important for plant survival. As part of the plant's secondary metabolites, essential oils attract pollinators, reduce abiotic stress, and protect plants from pests and herbivores, among other direct and indirect roles. Secondary metabolites of lower volatility are not removed from plant material when essential oils are produced, which mostly happens through distillation. These other secondary metabolites include flavonoids, catechols, phenolic compounds, and other nonvolatile bioactive substances that may be recovered from the plant material to become valuable constituents of products for human well-being. A circular economy operation aims to take full advantage of essential oils, nonvolatile secondary metabolites, and secondary metabolite-depleted biomass.

Figure 1 summarizes common processing options applied to the essential oil value chain within the circular economy approach. The first step is the retrieval of essential oil from plant material, due to its exposure to steam or mechanical compression (in the case of citrus). After a condensation step (in the case of steam distillation), the essential oil is separated by simple decantation from the aqueous phase, hydrosol,

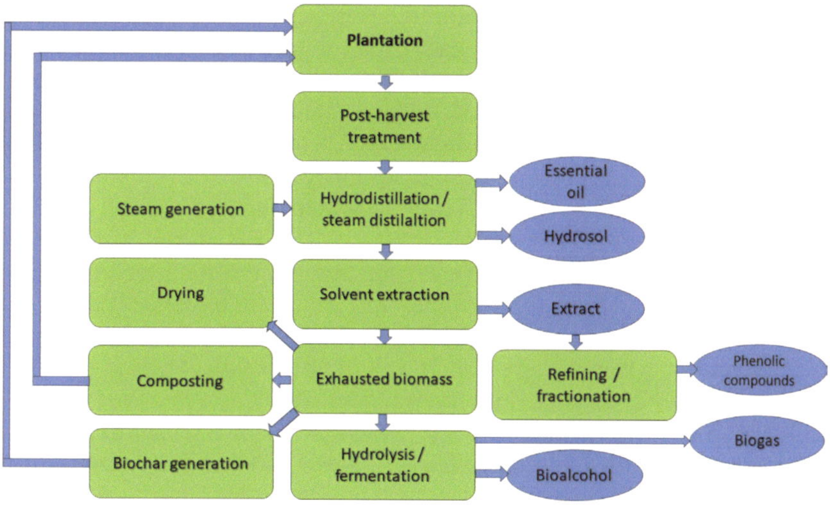

Figure 1.
Summary of the processing steps for the complete use of vegetal material in the essential oil agroindustry in observance of the circular economy principles.

which retains small amounts of some essential oil components that are not completely hydrophobic. The residual biomass enters a sequence of value-addition steps that have several potential final products and no waste. A very useful first task is the use of some extraction agent (ethanol, ethanol-water, CO_2) to remove waxes, pigments, flavonoids, and polyphenols from the residual biomass combined into a mixture known as the extract. This complex mixture may be further processed to obtain various fractions that are enriched in certain bioactive compounds. Composting, biochar generation, and biogas, or ethanol production from fermentation, are examples of several value-addition processes that may be applied to the lignocellulosic material that remains after solvent extraction. Thus, the circular economy version of the essential oil value chain has zero waste and several products, such as essential oil, hydrosol, extract, biofuel, biochar, and compost.

3. Hydrosol

Essential oils are decanted from the condensed water used as steam to separate them from the plant material. This condensed water, called hydrosol, hydrolat, distillation wastewater, or floral water, is obtained in amounts at least 50 times larger than the decanted essential oil and contains polar and hydrophilic essential oil components. A comparative study of the compositions of 44 hydrosols and their essential oils showed that in almost half of the hydrosols, the major component was different from that of the oil. Due to solubility differences, the concentrations of these organic substances are much smaller in hydrosol than in essential oil [6]. Some substances found in the hydrosol may result from molecular rearrangements caused by heating during distillation. For example, linalool and α-terpineol were found at higher concentrations in hydrosol than in *Lavandula angustifolia* essential oil [7].

There is a growing recognition of hydrosols as essential oil coproducts with many applications. *Mentha pulegium* and *Mentha suaveolens* hydrosols have shown high insecticidal effects against *Toxoptera aurantii*, a citrus pest [8]. A study on the control of the *Myzus persicae* aphid pest showed that the application of *Melissa officinalis* or *M. pulegium* hydrosols on eggplant leaves had an inhibitory effect, while the use of *Origanum marjorana* hydrosol caused 10–15% mortality after 24 h [9]. These and many more recent reports support the increased use of hydrosols in biological agriculture against mushrooms, mildew, and insects. The hydrosol from *Cuminum cyminum* seeds has caused decreased hatching of root-knot nematodes, a widely spread pest of many plantations [10].

Hydrosols are increasingly recognized as sources of natural ingredients for cosmetic, nutraceutical, and food applications. One important biological activity in this respect is antioxidant capacity. It was found that hydrosols from basil, sage, and rosemary wastes of packaged fresh aromatic plant production contained caffeic acid derivatives, glycosylated luteolin, and other flavonoids, whose presence was manifested in their strong antioxidant capacity, similar to that of pomegranate juice and higher than that of red wine and green tea [11]. Bactericidal activity is also important for certain applications. A study of the hydrosols of basil, cardamom, clove, cinnamon, and thyme showed that they produced inhibitory effects against *Salmonella typhi*, *Staphylococcus aureus*, and *Escherichia coli* [12].

4. Extract

The residual biomass from the distillation of aromatic plants to obtain essential oils contains valuable components such as carotenoids, carbohydrates, lipids, flavonoids, and phenolic acids within a lignocellulosic matrix. The current destiny of most distillation residues is a garbage dump. When this is not the case, common applications are the transformation into biofuel (as raw lignocellulosic bagasse, or with further processing into biochar) or as a major ingredient for composting operations. However, an important and profitable step may be inserted before these treatments to take advantage of the presence of flavonoids and other bioactive molecules in these residues. Extraction techniques may be employed to obtain fractions enriched in these bioactive substances. Maceration, Soxhlet extraction, microwave-assisted extraction, ultrasonic-assisted extraction, supercritical extraction techniques, pulse electric field extraction, enzyme-assisted extraction, molecular distillation, and accelerated solvent extraction are some of the tools employed to obtain extracts from plant materials. Further processing may involve ultrafiltration, nanofiltration, membrane filtration, supercritical antisolvent fractionation, or other physical processes that help to separate extract components according to molecular size. Rosemary (*Salvia rosmarinus*) is a medicinal and aromatic herb that contains bioactive compounds (carnosic acid, carnosol, rosmarinic acid) of interest in many fields. Many of the previously mentioned techniques have been applied to fresh and residual rosemary biomass and this collective effort permits the comparison of these techniques and their operating conditions according to yield, extract stability, costs, and energy demands [13].

The efficient recovery of bioactive compounds from the distillation residual biomass is an important subject of current research in the field of aromatic plants. A search of scientific publications from 2000 to July 2023 on "aromatic plant" in Scopus produced 28,973 results, more than 25% of which (6556) included the word "waste." Many works have determined the compositions of extracts obtained from the biowaste of individual aromatic species at the laboratory scale with various techniques. Polyphenols are frequent constituents, and their presence is related to the antioxidant or antimicrobial activities determined for the extracts. It is now clear that the solid wastes from essential oil distillation are rich sources of bioactive molecules with applications in the cosmetic, food, pharmaceutical, hygiene, and other industries. The challenges are scaling up the extraction processes and developing extract fractionation processes toward enrichment in specific sets of compounds or isolating individual substances. The following are examples of the general situation of the addition of value to aromatic plant distillation waste through the production of extracts.

The term "polyphenols" designates phenolic substances that contain several hydroxyl groups but does not imply that they have a polymeric nature. A more general denomination is "phenolic compounds." They include hydroxybenzoic acids, hydroxycinnamic acids, phenylpropanoids, coumarins, flavonoids, and phenolic terpenes [14].

Rosa alba waste from steam distillation and supercritical CO_2-extracted fresh flowers and steam distillation wastes from *L. angustifolia*, *M. officinalis*, and *Ocimum basilicum* essential oil production were dried at 50°C, ground, sieved (0.5 mm), and macerated with 70% ethanol (1:6 w/v) at 60°C (1 h) and room temperature (24 h). The CO_2-extracted *Rosa alba* afforded the extract with the highest polyphenol content (11 g/L), followed by melissa waste (6.6 g/L). Flavonoids such as rutin (1.2 g/L), catechin (1.1 g/L), and quercetin-3-glucoside (0.7 g/L) were quantified in these extracts using HPLC-DAD. Basil waste produced the extract with the highest rosmarinic acid content (1.2 g/L). Other phenolic acids, such as gallic, ferulic, and 3,4-dihydroxybenzoic acids, were found in the extracts at concentrations between 0.1 and 0.6 g/L. Thus, the distillation wastes from these aromatic plants are a rich source of polyphenols that could be used as supplements to increase antioxidant activities in food [15].

Ultrasonic agitation (37 kHz, 30°C) and 70% methanol were used to obtain extracts from residual distillation biomass of six aromatic species (*M. officinalis*, *Mentha spicata*, *Origanum vulgare*, *Salvia fruticosa*, *S. rosmarinus*, and *Satureja thymbra*). LC-MS analysis of these extracts identified a total of 48 compounds, including 20 phenolic acids, 26 flavonoids, and 2 phenolic diterpenes. Phenolic acids varied from 3817 mg/100 g (*S. rosmarinus*) to 14,462 mg/100 g (*M. spicata*). Flavonoids varied from 747 mg/100 g (*S. fruticosa*) to 3112 mg/100 g (*Satureja thymbra*) [16]. Rosmarinic acid is a frequent component of residual biomass extracts, determined at concentrations between 0.7 and 154 mg/g of extract [17]. Its higher concentrations have been reported in residual biomass extracts of *M. officinalis* (93.3 mg/g), *M. spicata* (96.6 mg/g), and *Thymus vulgaris* (105 mg/g) [18].

Residual rosemary hydrodistillation biomass was dried, ground, and extracted with ethanol under ultrasonic agitation. LC-MS analysis showed that 60% of the chromatographic area was represented by carnosol (35.6%), carnosic acid (12.1%), cirsimaritin (9.1%), and genkwanin (4.7%). The absence of rosmarinic acid was attributed to its thermal degradation evidenced by the presence of caffeic acid (0.9%) in the extract, and at its dissolution in the hydrosol [19]. The antioxidant activity of this extract was high, similar to that of the extract obtained from red grape pomace. It had insect antifeeding effects on *Leptinotarsa decemlineata* Say (Coleoptera:Chrysomelidae) (polyphagous/olyphagous chewing insects) and the aphid *Myzuspersicae sulzer* (Hemiptera:Aphididae).

The unfractionated rosemary extract can be used as an antioxidant or as a natural crop protectant, among many other applications. However, there are continuous efforts to isolate its main components or to obtain enriched fractions. Increased yields of ursolic acid (15.8 mg/g), rosmarinic acid (15.4 mg/g), and oleanolic acid

(12.2 mg/g) were obtained in ultrasound-assisted extraction by varying the pH, ethanol%, temperature, and solvent:solid ratio [20].

Thyme (*T. vulgaris*) is a common aromatic plant used in the traditional medicine, pharmaceutical, and food industries [21]. Residual biomass from thyme distillation was macerated with 75% ethanol (1:10 w/v) at room temperature for 1 day. The extract was obtained with a 3.85% yield with a phenolic acid content of 62 mg of gallic acid equivalents per gram of dry extract. LC-MS analysis revealed that its main components were rosmarinic acid (105 mg/g) and rutin (87 mg/g) [18].

Culinary herbs of many cultures include some types of oregano, of which there are several species from various origins. Carvacrol and thymol are the most common oregano essential oil components and are also the main contributors to the bioactive properties of this herb. The variability in origin and habitat is reflected in the reported 20-fold variation in carvacrol content found in comparisons of essential oils obtained from different oregano species [22]. Mexican oregano is mainly represented by *Lippia graveolens*, whose essential oil is rich in thymol in carvacrol. The analogous species in northern South America is *Lippia origanoides*, which in Brazil is more commonly recognized as *Lippia sidoides*, a synonym according to genetic studies [23]. Similar to other species, there are many *L. origanoides* and *L. graveolens* chemotypes, which are distinct populations within the same species with different secondary metabolite profiles. There are reports of at least five *L. origanoides* chemotypes, some of which have no thymol or carvacrol in their essential oils [24, 25].

One approach to the complete utilization of *L. origanoides* categorizes essential oil and hydrosol as products of the distillation process and directs the residual plant material to various purposes, such as extraction, composting, or combustion material for steam generation. A patent has been granted for this process, specifically applied to *L. origanoides* [26].

Extraction with ethanol-modified supercritical CO_2 of the residues from thymol-rich *L. origanoides* distillation afforded a resin that contained 20 g of flavonoids/kg. When applied to the phellandrene-rich *L. origanoides* chemotype, oleoresin contained 31 g of valuable pinocembrin/kg [27]. This relatively large content of pinocembrin motivated further studies on solubility [28] in CO_2, and mass transfer models [29]. A recent report of this work showed that the use of two coexisting fluid phases with various proportions of ethanol, water, and CO_2, afforded an extract containing 145 g pinocembrin/kg, which is approximately a fivefold increase relative to the concentration obtained with ethanol-modified CO_2 [30].

5. Soil amendments

The pyrolysis of biomass under low oxygen conditions produces biochar, which is a carbonaceous material with agricultural applications due to its porous surface and nutrient content. The estimated annual residual biomass from the distillation of *Mentha arvensis*, *Mentha citrata*, and *M. piperita* is 10.5 thousand million tons [31]. These wastes are burned or composted, but both approaches face implementation problems. The use of aromatic plant biomass has been associated with antigerminating attributes upon composting [32]. Proper burning to secure complete combustion requires a considerable initial investment. An alternative to burning and composting is a two-step sequential approach in which solvent treatment of the residual biomass is used to obtain an extract with antioxidant capacity, and the plant material is subsequently heated under an inert atmosphere to obtain a biochar. The biochar is used for soil amendment [33].

An interesting alternative approach that reduces investment and operating costs is to involve the distillation plants themselves in biochar production. The ash pit of a distillation plant with 500 g plant material capacity was used to hold stainless steel boxes (25 L) that contained biomass waste (10 kg) and was provided with eight holes for the release of volatile compounds. A residence time of 2 h was suitable to produce biochar in up to 60% yield and 0.26–1.6 g/m^3 density [34].

Crop cultivation leftovers from pruning and plant selection at a commercial production of fresh basil, rosemary, and sage were used for essential oil and hydrosol production and the distillation residue was employed in on-farm composting. Temperature monitoring showed that 30 days was sufficient to obtain usable compost [35].

6. Other end uses

Combustible bio-oil was obtained by pyrolysis of *Cymbopogon flexuosus* distillation wastes in a fixed-bed reactor. Due to the low lignin content of the biomass, the bio-oil had a lower content of phenolic compounds than those obtained from other biomasses. It also had low concentrations of polyaromatic hydrocarbons and nitrogenous compounds. Combustion heat in the order of 30 MJ/kg was measured for this bio-oil [36].

The use of lavender (*L. angustifolia* Miller) distillation residues in mushroom production was examined by cultivating *Pleorotus ostreatus* on substrates composed of shredded lavender waste and barley straw, maintained at 24°C and 65–70% humidity for 50–60 days. Once the mushrooms were removed, the exhausted substrate was used as a soil conditioner or to produce biocompost. The entire mushroom production process was established to be cost-effective [37].

7. Citrus

Citrus fruits constitute a separate case because their essential oils are normally obtained by mechanical compression of the peel, not by steam distillation. Citrus fruits are the largest fruit crop in the world (over 100 million tons per year) [38] and their main destiny is the food industry. After processing, the peels, seeds, and membrane residue represent approximately 60% of the original fresh fruit mass. Essential oil is present in the colored exterior part of the peel, called the flavedo. The main component of citrus essential oils is limonene (above 80%, depending on the species). Citrus essential oils have antioxidant, antidiabetic, insecticidal, antifungal, and antibacterial properties, which make them valuable ingredients for applications in the pharmaceutical, sanitary, cosmetic, agricultural, and food industries [39]. The inner walls of the citrus fruit peel (albedo) are rich (~30%) in pectin. Pectin is a mixture of acid and neutral-branched polysaccharides that contain glucuronic acid. Pectin is used as a thickener, emulsifier, gelling agent, or fat substitute in the food and beverage industry. Citrus peel also contains flavonoids for which antioxidant, anticancer, anti-inflammation, and cardiovascular protection activities have been reported [40]. Thus, processing citrus fruit residual material with a zero-waste approach may lead to value addition through the isolation of essential oil, pectin, flavonoids, and other bioactive molecules.

The most common primary approaches to treat citrus fruit waste involve essential oil separation, followed by the use of the residue for composting and for animal food. Some of the limiting factors to using citrus peel waste in composting are its low nitrogen content, and its antimicrobial properties (associated with limonene), which have a negative impact on soil microorganisms. Animal food is another potential use of citrus peel waste, but there have been mixed positive and negative experiences. A study in which broiler finisher birds were fed with increasing amounts of orange peel concluded that the inclusion of sweet orange peels had adverse effects on the growth rate and nutrient utilization by the birds [41]. This was attributed to the decrease in feed intake because the compounded diets became unpalatable upon the inclusion of sweet orange peels, some of them with certain fermentation levels. The processing of citrus peel waste through anaerobic digestion and consecutive fermentations may lead to the isolation of several bioactive products, but it requires the previous removal of limonene, which can inhibit microbial activity [42].

The essential oil in the peel of citrus fruits may be isolated by mechanical pressing (cold press), which is the most common method, although higher yields are obtained with hydrodistillation and steam distillation. The latter have higher costs due to the energy needed, but when complete biomass utilization is the goal, these techniques have the advantage that the residual biomass has low amounts of limonene and can be subjected to liquid fermentation without further processing. An interesting alternative is microwave-assisted distillation, which due to its shorter duration has lower total energy costs. In one of its implementations, the moisture content of the vegetal material is sufficient to absorb the supplied microwave energy, and no additional water is needed. This lowers operation costs and the volume of the treatment chamber required [43].

8. Conclusions

The examples presented outline the importance of embracing circular economy principles within the essential oil agroindustry to mitigate environmental problems and enhance sustainability. The substantial amounts of residual biomass generated during essential oil production can be transformed into valuable resources through innovative approaches and advanced technologies. The circular economy approach offers a solution to the environmental challenges posed by residual biomass from essential oil production. Advanced extraction techniques play an important role in recovering bioactive compounds from residual biomass. These compounds, such as flavonoids and phenolic acids, have diverse applications in industries such as cosmetics, pharmaceuticals, and food. Hydrosols, byproducts of essential oil distillation, have emerged as valuable coproducts because their insecticidal, antioxidant, and antibacterial properties make them versatile assets in various sectors. Circular economy principles extend to the utilization of residual biomass for producing biochar and compost. These products have applications in soil conditioning, agriculture, and energy generation, contributing to waste reduction and resource efficiency.

In conclusion, the integration of circular economy principles in the essential oil agroindustry holds immense potential to alleviate environmental concerns and contribute to sustainable development. By extracting maximum value from residual biomass, the industry can create a more environmentally friendly and economically viable production process. Through innovative techniques, collaboration across sectors, and ongoing research, the vision of circular and sustainable essential oil agroindustry can become a reality.

Acknowledgements

The authors are thankful for the funding provided by Ministerio de Ciencia, Tecnología e Innovación, Ministerio de Educación Nacional, Ministerio de Industria, Comercio y Turismo, and ICETEX, Programm Ecosistema Científico-Colombia Científica from Fondo Francisco José de Caldas, Grant RC-FP44842-212-2018, and the General Royalties System, BPIN-2018000100044.

Conflict of interest

The authors declare no conflict of interest.

Author details

Elena Stashenko* and Jairo René Martínez
Research Center for Biomolecules, Industrial University of Santander, Bucaramanga, Colombia

*Address all correspondence to: elena@tucan.uis.edu.co

References

[1] World Trade Center. www.intracen.org Consulted 11/3/2023

[2] Barreiro-Gen M, Lozano R. How circular is the circular economy? Analysing the implementation of circular economy in organisations. Business Strategy and the Environment. 2020;**29**:3484-3494. DOI: 10.1002/bse.2590

[3] Bhawna B, Prateek G, Junaid A, Vlaskin MS, Kurbatova A, Adarchenko I, et al. A review on holistic approaches for fruits and vegetables biowastes valorization. Materials Today: Proceedings. 2023;**73**:54-63. DOI: 10.1016/j.matpr.2022.09.168

[4] Machado BA, Pereira CG, Nunes SB, Padilha FF, Umsza-Guez MA. Supercritical fluid extraction using CO_2: Main applications and future perspectives. Separation Science and Technology. 2013;**48**:2741-2760. DOI: 10.1080/01496395.2013.811422

[5] Cheng SY, Tan X, Show PL, Rambabu K, Banat F, Veeramuthu A, et al. Incorporating biowaste into circular bioeconomy: A critical review of current trend and scaling up feasibility. Environmental Technology and Innovation. 2020;**19**:101034. DOI: 10.1016/j.eti.2020.101034

[6] Inouye S, Takahashi M, Abe S. A comparative study on the composition of forty four hydrosols and their essential oils. International Journal of Essential Oil Therapeutics. 2008;**2**:89-104

[7] Baydar H, Kineci S. Scent composition of essential oil, concrete, absolute and hydrosol from lavandin (LavandulaxintermediaEmeric ex Loisel.). Journal of Essential Oil-Bearing Plants. 2009;**12**:131-136. DOI: 10.1080/0972060X.2009.10643702

[8] Zekri N, Handaq N, El Caidi A, Zair T, El Belghiti MA. Insecticidal effect of Mentha pulegium L. and Mentha suaveolens Ehrh. hydrosols against a pest of citrus, Toxoptera aurantii (Aphididae). Research on Chemical Intermediates. 2016;**42**:1639-1649. DOI: 10.1007/s11164-015-2108-0

[9] Petrakis EA, Kimbaris AC, Lykouressis DP, Polissiou MG, Perdikis D. Hydrosols evaluation in pest control: insecticidal and settling inhibition potential against Myzus persicae (Sulzer). Journal of Applied Entomology. 2015;**139**:260-267. DOI: 10.1111/jen.12176

[10] Pardavella I, Daferera D, Tselios T, Skiada P, Giannakou I. The use of essential oil and hydrosol extracted from cuminum cyminum seeds for the control of Meloidogyne incognita and Meloidogyne javanica. Plants. 2021;**10**:46. DOI: 10.3390/plants10010046

[11] Celano R, Piccinelli AL, Pagano I, Roscigno G, Campone L, De Falco E, et al. Oil distillation wastewaters from aromatic herbs as new natural source of antioxidant compounds. Food Research International. 2017;**99**:298-307. DOI: 10.1016/j.foodres.2017.05.036

[12] Hussien J, Teshale C, Mohammed J. Assessment of the antimicrobial effects of some Ethiopian aromatic spice and herb hydrosols. International Journal of Pharmacology. 2011;**7**:635-640. DOI: 10.3923/ijp.2011.635.640

[13] Ali A, Chua BL, Chow YH. An insight into the extraction and fractionation technologies of the essential oils and

bioactive compounds in Rosmarinus officinalis L.: Past, present and future. TrAC Trends in Analytical Chemistry. 2019;**118**:338-351. DOI: 10.1016/j.trac.2019.05.040

[14] Tsimogiannis D, Oreopoulou V, Watson RR. Classification of phenolic compounds in plants. In: Watson R, editor. Polyphenols in Plants. 2nd ed. London: Academic Press; 2019. pp. 263-284. DOI: 10.1016/B978-0-12-813768-0.99926-8

[15] Slavov A, Vasileva I, Denev P, Dinkova R, Teneva D, Ognyanov M, et al. Polyphenol-rich extracts from essential oil industry wastes. Bulgarian Chemical Communications. 2020;**52**:78-83

[16] Irakli M, Skendi A, Bouloumpasi E, Chatzopoulou P, Biliaderis CG. LC-MS identification and quantification of phenolic compounds in solid residues from the essential oil industry. Antioxidants. 2021;**10**:2016-2035. DOI: 10.3390/antiox10122016

[17] Skendi A, Irakli M, Chatzopoulou P, Bouloumpasi E, Biliaderis CG. Phenolic extracts from solid wastes of the aromatic plant essential oil industry: Potential uses in food applications. Food Chemistry Advances. 2022;**1**:100065. DOI: 10.1016/j.focha.2022.100065

[18] Gavarić N, Kladar N, Mišan A, Nikolić A, Samojlik I, Mimica-Dukić N, et al. Postdistillation waste material of thyme (Thymus vulgaris L., Lamiaceae) as a potential source of biologically active compounds. Industrial Crops and Products. 2015;**74**:457-464. DOI: 10.1016/j.indcrop.2015.05.070

[19] Santana-Méridas O, Polissiou M, Izquierdo-Melero ME, Astraka K, Tarantilis PA, Herraiz-Peñalver D, et al. Polyphenol composition, antioxidant and bioplaguicide activities of the solid residue from hydrodistillation of Rosmarinus officinalis L. Industrial Crops and Products. 2014;**59**:125-134. DOI: 10.1016/j.indcrop.2014.05.008

[20] Bernatoniene J, Cizauskaite U, Ivanauskas L, Jakstas V, Kalveniene Z, Kopustinskiene DM. Novel approaches to optimize extraction processes of ursolic, oleanolic and rosmarinic acids from Rosmarinus officinalis leaves. Industrial Crops and Products. 2016;**84**:72-79. DOI: 10.1016/j.indcrop.2016.01.031

[21] Burt S. Essential oils: their antibacterial properties and potential applications in foods – a review. International Journal of Food Microbiology. 2004;**94**:223-253. DOI: 10.1016/j.ijfoodmicro.2004.03.022

[22] Fleisher A, Sneer N. Oregano spices and origanum chemotypes. Journal of the Science of Food and Agriculture. 1982;**33**:441-446. DOI: 10.1002/jsfa.2740330508

[23] Almeida MC, Pina E, Hernandes C, Zingaretti S, Taleb-Contini S, Salimena F, et al. Genetic diversity and chemical variability of Lippia spp. (Verbenaceae). BMC Research Notes. 2018;**11**:725. DOI: 10.1186/s13104-018-3839-y

[24] da Silva AP, Silva NF, Andrade EHA, Gratieri T, Setzer WN, Maia JGS, et al. Tyrosinase inhibitory activity, molecular docking studies and antioxidant potential of chemotypes of Lippia origanoides (Verbenaceae) essential oils. PLoS One. 2017;**12**:e0175598. DOI: 10.1371/journal.pone.0175598

[25] Ribeiro AF, Andrade EH, Salimena FR, Maia JG. Circadian and seasonal study of the cinnamate chemotype from Lippia origanoides Kunth. Biochemical Systematics and Ecology. 2014;**55**:249-259. DOI: 10.1016/j.bse.2014.03.014

[26] Stashenko E, Martínez JR, Durán DC, Arias AJ, Córdoba Y, Mejía J, et al. Patent of invention US11028338B2: Method for making full use of Lippia origanoides. 2019

[27] Arias J, Mejía J, Córdoba Y, Martínez JR, Stashenko E, del Valle JM. Optimization of flavonoids extraction from Lippia graveolens and Lippia origanoides chemotypes with ethanol-modified supercritical CO_2 after steam distillation. Industrial Crops and Products. 2020;**146**:112170. DOI: 10.1016/j.indcrop.2020.112170

[28] Arias J, Martínez J, Stashenko E, del Valle JM, de la Fuente JC. Supercritical CO_2 extraction of pinocembrin from Lippia origanoides distillation residues. 1. Multicomponent solubility and equilibrium partition. Journal of Supercritical Fluids. 2022;**180**:105426. DOI: 10.1016/j.supflu.2021.105426

[29] Arias J, Martínez J, Stashenko E, del Valle JM, Núñez G. Supercritical CO_2 extraction of pinocembrin from Lippia origanoides distillation residues. 2. Mathematical modeling of mass transfer kinetics as a function of substrate pretreatment. Journal of Supercritical Fluids. 2022;**180**:105458. DOI: 10.1016/j.supflu.2021.105458

[30] Arias J, Muñoz F, Mejía J, Kumar A, Villa AL, Martínez JR, et al. Simultaneous extraction with two phases (modified supercritical CO_2 and CO_2-expanded liquid) to enhance sustainable extraction/isolation of pinocembrin from Lippia origanoides (Verbenaceae). Advances in Sample Preparation. 2023;**6**:100059. DOI: 10.1016/j.sampre.2023.100059

[31] Rawat AP, Kumar V, Singh DP. A combined effect of adsorption and reduction potential of biochar derived from Mentha plant waste on removal of methylene blue dye from aqueous solution. Separation Science and Technology. 2020;**55**:907-921. DOI: 10.1080/01496395.2019.1580732

[32] Martino LD, Mancini E, Almeida LF, Feo VD. The antigerminative activity of twenty-seven monoterpenes. Molecules. 2010;**15**:6630-6637. DOI: 10.3390/molecules15096630

[33] Saha A, Basak BB, Banerjee A. In-vitro antioxidant evaluation and production of biochar from distillation waste biomass of Mentha arvensis. Journal of Applied Research on Medicinal and Aromatic Plants. 2022;**31**:100428. DOI: 10.1016/j.jarmap.2022.100428

[34] Ahsan M, Singh M, Singh RP, Yadav V, Tandon S, Saikia BK, et al. An innovative circular model for recycling the wastes into biochar using distillation units. Journal of Cleaner Production. 2022;**361**:132258. DOI: 10.1016/j.jclepro.2022.132258

[35] Zaccardelli M, Roscigno G, Pane C, Celano G, Di Matteo M, Mainente M, et al. Essential oils and quality composts sourced by recycling vegetable residues from the aromatic plant supply chain. Industrial Crops and Products. 2021;**162**:113255. DOI: 10.1016/j.indcrop.2021.113255

[36] Deshmukh Y, Yadav V, Nigam N, Yadav A, Khare P. Quality of bio-oil by pyrolysis of distilled spent of Cymbopogon flexuosus. Journal of Analytical and Applied Pyrolysis. 2015;**115**:43-50. DOI: 10.1016/j.jaap.2015.07.003

[37] Di Piazza S, Benvenuti M, Damonte G, Cecchi G, Mariotti MG, Zotti M. Fungi and circular economy: Pleurotus ostreatus grown on a substrate with agricultural waste of lavender, and its promising biochemical profile.

Recycling. 2021;**6**:40. DOI: 10.3390/recycling6020040

[38] USDA. Citrus: World markets and trade. 2023. Available from: https://www.fas.usda.gov/data/citrus-world-markets-and-trade [Accessed 10 March, 2023]

[39] Eristanna P, Vito AL, Maria AG. Current and potential use of citrus essential oils. Current Organic Chemistry. 2013;**17**:24. DOI: 10.2174/13852728113179990122

[40] Addi M, Elbouzidi A, Abid M, Tungmunnithum D, Elamrani A, Hano C. An overview of bioactive flavonoids from citrus fruits. Applied Sciences. 2021;**12**:29. DOI: 10.3390/app12010029

[41] Ani AO, Iloh EA, Akinsola OO. Dietary effect of processed orange peels on growth performance of broiler finisher birds. Current Journal of Applied Science and Technology. 2015;**9**:576-583. DOI: 10.9734/BJAST/2015/6052

[42] Calabrò PS, Paone E, Komilis D. Strategies for the sustainable management of orange peel waste through anaerobic digestion. Journal of Environmental Management. 2018;**212**:462-468. DOI: 10.1016/j.jenvman.2018.02.039

[43] Ferhat MA, Boukhatem MN, Hazzit M, Meklati BY, Chemat F. Cold pressing, hydrodistillation and microwave dry distillation of citrus essential oil from Algeria: A comparative study. Electronic Journal of Biology. 2016;**1**:30-41

Chapter 11

Essential Oils Based Nano Formulations against Postharvest Fungal Rots

Gull-e-laala Khan, Gulshan Irshad, Raina Ijaz, Nagina Rafiq, Sajid Mehmood, Muhammad Usman Raja, Abd-ur-Rehman Khalid, Farah Naz, Nayla Haneef, Saiqa Bashir, Amna Maqsood and Amar Mehmood

Abstract

Postharvest phytopathogenic rot fungi affect the quality and quantity of perishable fruits and vegetables. About 30–40% peaches deteriorate annually after harvest in world whereas 40–50% losses are reported from Pakistan. Our research envisages importance of an eco-friendly plant essential oils based nano formulations as a management strategy against postharvest deteriorating fungal rots by enhancing their shelf-life and to attenuate reliance on synthetic fungicides. Plant essential oils mode of action against fungal postharvest rots is responsible of rupturing plasma membrane of fungal cell wall. The natural ripening process of perishable commodities does not get affected by the presence of antifungal packaging in the form of plant essential oil nano formulations as no significant alteration in weight loss of produce was recorded. Challenges in applying EOs for microbial suppression in postharvest systems include optimizing their positioning in commercial fruit storage containers. Several innovative approaches are analyzed in terms of work environment and implementation regarding disease management along with future perspectives in concerning field.

Keywords: fungal plant pathology, postharvest pathology, bio management of fungal rots with plant essential oils, nanoencapsulation, molecular phylogenetics

1. Introduction

Postharvest fruit losses usually develop along the food supply chain by fungal rots during handling, storage, transportation, processing and marketing, thus resulting in the decline of quantity, quality and shelf life of perishable fruits. Fungal postharvest rots pose a serious threat limiting the market value of peaches, resulting in serious economic losses worldwide [1]. Furthermore, postharvest fungal rots are often the major concern influencing consumer requirements. Accordingly, postharvest deteriorations primarily progress from damages that arise most prominently before, during and after harvesting. Once spores of fungi are inoculated in fruit from these wounds, they result in rapid fruit deterioration. Germinating conidia of fungi may enter intact fruit cuticle and establish internally in the host [2]. These infections completely develop into active decay as fruit is fully matured and become susceptible.

The high metabolic activity of peaches results in ethylene production and is mainly subjected to rapid quality decrease. This causes several negative effects *viz*. increased fungal rots susceptibility and fruit physiological disorders consequently resulting in the growth potential of microbes on fruit surface [3]. Various postharvest fungal rots of the genera; *Fusarium, Penicillium, Alternaria, Botrytis, Cladosporium, Colletotrichum, Trichothecium, Aspergillus* and Stigmina deteriorate quality of the stone fruits [4, 5]. To reduce conservational pollution owing to fungicides their excessive use against phytopathogenic rots is reduced in the previous years. Moreover, biological control has emerged as an applicable strategy to combat major fungal postharvest deterioration of perishable fruits [6, 7].

2. Postharvest losses

Postharvest fruit losses mostly arise laterally in the entire food supply chain from handling to consumption declining quality and market value of produce [1]. Usually, immature peaches show no visible rot symptoms but as they mature symptoms are clearly visible. These latent infections usually become active when the fruit ripen, worsening the disease incidence in harvest and post-harvest. *Fusarium* spp., *Rhizopus* spp., *Penicillium* spp., *Cladosporium* spp., *Botrytis* spp., *Aspergillus* spp., *Monilinia* spp., are major destructive postharvest fungal rots infecting perishable stone fruits like peaches. These postharvest fungal rots are now frequently reported in regions with high technology stores resulting in fruit rotting, and it is estimated that about 25–50% of the produce is lost in the developing and developed countries, respectively nectarines [8].

Postharvest diseases of stone fruits are characteristically triggered by a diversity of fungal rots, yet some of their casual agents, confirmation is still progressing. Sufficient awareness regarding epidemiology of opportunistic fungal rots is available whereas knowledge of their occurrence after quiescent infections during long term storage is limited. Usually, storage rots cause significant losses after latent infections [9]. Fungal pathogens cause significant fruit losses after harvest and results consequently in an essential reduction in the global supply. Several strategies have been altered to manage postharvest fungal rots around the globe. Fungal Rots caused by phytopathogenic postharvest rots incite severe agricultural and horticultural crop losses annually [10]. Bio-management with antimicrobial agents is explored as a substitute against postharvest fungal rots of perishable fruits [11].

3. Hazardous affects of synthetic fungicides

Fungicide use is very common at postharvest stage but as far as fruits are concerned, use of fungicide after harvest is strongly prohibited by FAO. The applications of synthetic fungicides rapidly increased throughout the preceding era. About four hundred thousand tons of fungicides are globally applied, which represents 17.5% of universal pesticide applications. The practice of hazardous fungicides to manage phytopathogenic postharvest fungal rots is limited, due to the toxic residual effect and long degradation period.

Application of synthetic fungicides against postharvest rots is known to be vastly effective and widely practical technique in orchards [12]. On the other hand, these chemicals have toxicologic hazards, which are dangerous to human health and cause ecological pollution. It is pertinent to mention that in few developed countries the use of fungicides is strongly prohibited by law after harvesting. Presently, the trend of using synthetic fungicides is reducing and practice of their use in agricultural products is rapidly minimizing [13]. Long term extensive practices of these hazardous fungicides may create resistance in pathogens, leading to severe conservational pollution with a serious threat to human health [14].

Lately, fludioxonil and azoxystrobin stayed registered in the USA as a post-harvest management application against peach decaying rots. Still, post-harvest use of these fungicides in European Union are banned because of fungicide regulatory concerns. Moreover, community stresses towards decrease in fungicides application, encouraged by more awareness of conservational and health concerns. The lack of an effective post-harvest strategy against fungal rot of peaches focuses the need for developing novel and eco-friendly control methods [13].

4. Plant essential oils

Fungicidal action of various plant EO's have been widely examined under various environmental conditions [15, 16]. Essential oils such as fenugreek, eucalyptus, turmeric, lime, thyme, fennel, clove, sage and peppermint are reported for their suppressive effect against postharvest fungal peach fruit rots [17]. Different techniques to apply essential oils against these decaying rots have been reported including; dipping, *in situ* uses and spraying however, there are several interests above potential sensory contamination concerns [18].

Use of synthetic fungicides as a chemical control is widely applied and highly effective method after harvesting fruits and vegetables [12]. However, it is pertinent to mention that fungicides have toxicological risks effecting environment and human health. World is moving towards trend of minimizing abundant use of synthetic chemicals in agricultural products [13]. Long term extensive practices of chemical fungicides may result in pathogen resistance, leading to severe conservational pollution posing hazardous effects on health [14].

An excessive concern regarding essential oils and plant extracts application as a promising biological substitute of conservative synthetic fungicides is raised. This may be credited to environmental effluence and fungicide challenge in postharvest phytopathogenic rots [19]. Applications of plant EO boons an alternate eco-friendly strategy against postharvest fungal rots of fruits besides vegetable [20]. The biotic action of plant EO's as antifungal and antimicrobial agents is in debate. As a complex mix of various aromatic volatile compounds plant essential oils from numerous plant parts *viz.* leaves, flowers, cloves, rhizomes, buds, roots, etc. are consistently in practice against several postharvest fungal rots. Essential oils extracted from plants are widely in use due to their antimicrobial, antifungal, insecticidal, and cytotoxic properties [21]. Moreover, the eco-friendly use of plant essential oils is popular to increase the shelf life of food products, as consumers are always conscious about various health concerns raised by hazardous additives. Essential oils also symbolize a defense mechanism against postharvest rots owing to their antimicrobial characteristics [22].

Eco-friendly use of plant EO's against contagious pathogens present in many horticultural produces helps in analysis of positive outcomes towards efficient options regarding fruit protection. Plant essential oils from various natural herbs and plants possess strong antifungal characteristics, *in-vitro*, that it may be used as a natural and eco-friendly strategy against perishable fruits decaying [18]. Stone fruits are highly sensitive to fungal infections. Currently, the management of postharvest fruit rots with plant essential oils (EOs) has been considerably observed as an innovative trend in biological conservation.

Plant Essential oils (PEOs) as volatile liquids obtained by several extraction techniques from plants and herbs, hold plentiful natural bioactive compounds with anti-fungal characteristics. Several studies regarding fungicidal characteristics of plant essential oils against postharvest fungal rots were reported *viz. Penicillium* spp., *Fusarium* spp., *Cladosporium* spp., *Alternaria* spp., *Aspergillus* spp. and *Botrytis cinerea* [23–25].

5. Plant essential oils based nano formulations against deteriorating fungal postharvest rots

Postharvest fruit decay is controlled with essential oils and is observed as an innovative drift in natural conservation [24]. Clove essential oil (CEO), as a natural antimicrobial agent, is generally recognized as a safe substance, shows strong in vitro activity against fungal rots [26]. Various studies revealed importance of essential oils *viz.* Eucalyptus [27], Thyme, Savory, Cinnamomum, Peppermint [28] and their antifungal efficacy against postharvest fungal rots of fruits *viz.* peaches, pears, apples, banana [29], citrus [30], strawberry, grapes, avocado, mango, and papaya. Another study revealed that organic extracts of various parts of *Lawsonia inermis* L. against *Aspergillus niger*, *Penicillium notatum*, *Fusarium oxysporum*, *Colletotrichum gloeosporioides* and *Rhizopus stolonifera* were highly effective in different environments [31, 32].

Furthermore, it was also reported that plant EO treatments have the prospective to control black, green and blue mold disease on fruits. It was narrated that the mycelial growth and conidial sprouting were distinctly affected by essential oils in a dose dependent manner [33]. Ultimately plant pathologists and scientists found excellent substitutes against postharvest fungal rots of perishable fruits and vegetables in the form of biocides [34].

Plant essential oils were applied against peaches with soft watery lesions and fluffy mycelium of fungus on outer surface, collected during fruit market survey at Cairo Egypt with soft watery lesions and fluffy mycelium of fungus on outer surface during fruit market and essential oils resulted insignificant reduction in growth of fungal mycelia [35].

However, the efficacy of plant essential oils holds significant importance in restraining pathogens mode of dispersal, by minimizing the spore load on fruit surfaces in the storage atmosphere [36]. Usually, plant essential oils are harmless together for the ecosystem and anthropological wellbeing, hence attention in their use as antifungal agents of postharvest fungal rots is increasing rapidly. Plant essential oils are non-hazardous to the treated fruits, vegetables and environment followed by human consumption. Plant EO's are elementary active natural pesticides and beneficial agents against various fungal phytopathogenic rots [37].

Plant EO's extracted from anise (*Pimpinella anisum* L.), thyme (*Thymus capitatus* L.), lemon (*Backhousia citriodora* F. Muell.) spearmint (*Menta spicata* L.) hold excellent antimicrobial nature and revealed inhibitory results against postharvest fungal peach rots. The fundamental component, citral, in citrus essential oil showed fungitoxic effect against various threatening postharvest phytopathogenic rots. Clove (*Syzgium aromaticum*) essential oil has been used since ages for its excellent properties including antifungal applications. In the previous dual eras, growing concern has been determined on clove EO to reduce postharvest fungal rots and progress storage quality of agricultural products [24, 26].

Moreover, eucalyptus (*Eucalyptus globulus*) essential oil possesses broad biotic activity, including anti-fungal, anti-microbial, acaricidal and nematicide features. The central components present in Eucalyptus EO, including eucalyptol, γ-terpinene, limonene, *p*-cymene, 1,8-cineole, α-pinene, ocimene α-terpineol, camphene and linalool play major fundamental role in disrupting cell wall of phytopathogenic rot fungi. Plant EO's are of utmost importance in holding fungistatic properties. Destruction of fungal spore germination by plant essential oils show fundamental role in limiting the spread of phytopathogenic rot pathogens [36]. Many essential oils can enhance shelf life and preserve quality harvested fruit while having no detrimental impact on the fruit itself [37].

Essential oils provide an alternative and more ecologically friendly strategy against fungal postharvest rots [20]. The fungicidal nature of plant essential oils might be owed to synergism amongst their mechanisms, subsequently, maximum activity has been reported to be improved mutually. The fungicidal action of plant EO's might be boosted by the technique of application. The possible potentially applicable strategies of with plant essential oils by immersion or spurting against postharvest phytopathogenic rot fungi has already been observed in fruit and vegetables. The pattern of several postharvest handlings may advance the efficacy of controlling postharvest fungal rots [38].

6. Antimicrobial packaging using plant essential oils

Challenges in applying EOs for microbial suppression in postharvest systems include optimizing their positioning in commercial fruit storage containers. One approach has been to dispense them from sachets (absorptive pads saturated with the plant EO) in the central of the packaging or enclosed towards it. Antimicrobic packaging in the form of antifungal sachet comprising of volatile constituents enclosed in them are the primary example of profitable packaging and are broadly used. Bioactivity of plant EO's vapor phase was known as a distinctive practice making them eye-catching against postharvest fungal rots in storage commodities [39].

To enhance the shelf life and promote market value of perishable fruits antimicrobial active packaging is of utmost importance. As it allows packaging material to interact directly with fruits for improving its quality [40]. Usually by adding a chemical compound in a deliberate way with specific antifungal ability into the packaging. Two major mechanisms may be applied in this regard, *viz.* internally in the packaging material during manufacturing, secondly in the headspace of the packaging as antimicrobial sachet during packaging [41, 42].

Other than antimicrobial packaging against postharvest fungal rots of perishable fruits it is pertinent to mention that fumigation with high concentration compounds was also applied as a postharvest treatment, where results showed water spots on external surface with browning and softening resulting in physiological damage. Many researchers argued that the same molecules at high concentrations may have adverse effects on climacteric fruit like peaches, leading to cell necrosis, and resulting in severe losses respectively [43]. Postharvest fungal rots *viz. F. semitectum, A. flavus,*

A. alternata, F. semitectum, L. theobromae, and *R. stolonifer* were controlled completely by using antifungal essential oil sachet.

The efficacy of EO's directly depend on the level of concentrations with significant alterations in dose dependent manner. Whereas the pathway of EO's mode of action against fungal postharvest rots might be related to their general capability of softening and dislocating the consistency of cell wall at various incubation temperatures [44].

7. Mechanism of antifungal constituents of plant essential oils

The mechanism of antifungal constituents in plant EO's *viz.* eugenol, eucalyptol & thymol include inhibition of enzymes by oxidization. Their possible concern is to damage membrane integrity and fungal cell wall degradation, which could result possible reduction in infections by post-harvest fungal rots [45]. Plant EO's as an excellent non-hazardous, antifungal agents may be adopted as an alternate management strategy to keep perishable fruits fresh during storage. Antifungal components eugenol in clove essential oil holds excellent inhibitory effect against postharvest storage fungal rots of perishable fruits [46]. One of the important factors concerning postharvest rots management is temperature in maintaining fruit quality prolonged shelf life [47].

Essential oils of various herbs and plants hold no significant impact in reducing weight of fruits during cold storage environment as compared to room incubation. It was reported that after application of EOs encapsulated with inclusion complexes average weight of stone fruits (nectarines) was not significantly reduced in cold storage [48]. Similarly, vapor treatment of *Thymus vulgaris* EO against fungal rots of peach fruit results minimum reduction of weight in cold room [49]. Cinnamon EO was also reported with a minimum reduction in average weight of peaches in cold storage incubation whereas in control (with no treatment) significant weight loss was recorded [50]. Other horticultural produce in cold storage treatments were also observed with minimum weight reduction after applying treatments of (carvacrol, eugenol, thymol, menthol, eucalyptol, oregano, and cinnamon) EO's whereas control fruit was observed with significant alterations in weight loss assessment [48, 51].

Plant essential oils mode of action against fungal postharvest rots is responsible of rupturing plasma membrane of fungal cell wall [52]. Volatile compounds in EO's acts by blocking ergosterol synthesis, internal leakage of fungal cell wall, disintegrating mycelium, disruption of cytoplasmic formation and ultimately triggering towards completed distortion [53–55].

Author details

Gull-e-laala Khan[1*], Gulshan Irshad[2], Raina Ijaz[3], Nagina Rafiq[4], Sajid Mehmood[2], Muhammad Usman Raja[2], Abd-ur-Rehman Khalid[1], Farah Naz[2], Nayla Haneef[1], Saiqa Bashir[3], Amna Maqsood[5] and Amar Mehmood[2]

1 Department of Plant Pathology, University of Poonch Rawalakot, Pakistan

2 Department of Plant Pathology, PMAS-Arid Agriculture University Rawalpindi, Rawalpindi, Pakistan

3 Department of Horticulture, University of Poonch, Rawalakot, Pakistan

4 Department of Food Science and Technology University of Poonch, Rawalakot, Pakistan

5 Department of Soil and Environmental Sciences, University of Poonch, Rawalakot, Pakistan

*Address all correspondence to: gulelalah.khan@yahoo.com; gulelalahkhan@upr.edu.pk

References

[1] Parfitt J, Barthel M, Macnaughton S. Food waste within food supply chains: Quantification and potential for change to 2050. Philosophy Transitional Research Society. 2010;**365**:3065-3081

[2] Adaskaveg JE, Forster H, Thompson DF. Identification and etiology of visible quiescent infections of *Monilinia fructicola* and *Botrytis cinerea* in sweet cherry fruit. Plant Disease. 2000:328-333

[3] Liu H, Cao J, Jiang W. Changes in phenolics and antioxidant property of peach fruit during ripening and responses to 1-methylcyclopropene. Postharvest Biology and Technology. 2015;**108**:111-118

[4] Snowdon AL. Color atlas of postharvest-harvest diseases and disorders of fruits and vegetables. In: General Introduction and Fruits. Vol. 1. London: Manson Publishing; 2010. 232 pp

[5] Hong CX, Michailides TJ. Prune, plum, and nectarine as hosts of Trichothecium roseum in California orchards. Plant Disease. 1997;**81**:112-121

[6] Zhang HY, Yang QY, Ge LL, Zhang GC, Zhang XL, Zhang XY. Chitin enhances biocontrol of Rhodotorula mucilaginosa to postharvest decay of peaches. International Journal of Biology and Macromolecules. 2016;**88**:465-475

[7] Janisiewicz WJ, Korsten L. Biological control of postharvest diseases of fruits. Annual Review Phytopathology. 2004;**40**:411-441

[8] Navarro D, Dıaz-Mula HM, Guillen F, Zapata PJ, Castillo S, Serrano M, et al. Reduction of nectarine decay caused by Rhizopus stolonifer, Botrytis cinerea and Penicillium digitatum with Aloe vera gel alone or with the addition of thymol. International Journal of Food Microbiology. 2011;**151**:241-246

[9] Marcel W, Bart PHJ. Latent postharvest pathogens of pome fruit and their management: From single measures to a systems intervention approach. European Journal of Plant Pathology. 2020;**156**:663-681

[10] Salman MAM. Biological Control of Rhizopus Soft Rot on Apple, Pear and Peach by Trichoderma Harzianum [Doctoral Thesis]. India: National University; 2005

[11] Zhang S, Zheng Q, Xu B, Liu J. Identification of the fungal pathogens of postharvest disease on peach fruits and the control mechanisms of Bacillus subtilis JK-14. Toxins. 2019;**11**:322

[12] Sansone G, Rezza I, Calvente V, Benuzzi D, Tosetti MIS. Control of Botrytis cinerea strains resistant to iprodione in apple with rhodotorulic acid and yeasts. Postharvest Biology and Technology. 2005;**35**:245-251

[13] Zheng Y, Jia X, Ran Y, Meijun D, Zhao Z, Chen L, et al. Inhibition effect of *aspergillus Niger* and quality preservation of apple by in-package sterilization medium flow of circulating. Scientia Horticullturae. 2022;**293**. DOI: 10.1016/j.scienta.2021.110708

[14] Shu C, Zhao H, Jiao W, Liu B, Cao J. Antifungal efficacy of ursolic acid in control of *Alternaria alternata* causing black spot rot on apple fruit and possible mechanisms involved. Scientia Horticultureae. 2019;**256**:231-252

[15] Chouhan SK, Sharma S, Guleria A. Antimicrobial activity of

some essential oils-present status and future perspectives. Medicines Basel. 2017;**4**(3):233-261

[16] Sivakumar DS, Bautista B. A review on the use of essential oils for postharvest decay control and maintenance of fruit quality during storage. Crop Protection. 2014;**64**:27-37

[17] Bosquez ME, Ronquillo E, Jesús S, Bautista BJ, Verdealvo J, Morales-López. Inhibitory effect of essential oils against Colletotrichum gloeosporioides and Rhizopus stolonifer in stored papaya fruit and their possible application in coatings. Postharvest Biology and Technology. 2010;**57**(2):132-137

[18] Antunes MDC, Cavaco AM. The use of essential oils for postharvest decay control. A review. Flavour and Fragrance Journal. 2010;**25**(5):351-366

[19] Mancini E, Camele I, Elshafie HS, De Martino L, Pellegrino C, Grulova D, et al. Chemical composition and biological activity of the essential oil of Origanum vulgare ssp. hirtum from different areas in the southern Apennines (Italy). Chemistry & Biodiversity. 2014;**11**(4):639-651

[20] Hammer KA, Carson CF, Riley TV. Antimicrobial activity of essential oils and other plant extracts. Journal of Applied Microbiology. 1999;**86**:985-990

[21] Faleiro L, Miguel GM, Guerrero CAC, Brito JMC. Antimicrobial activity of essential oils of Rosmarinus officinalis L., *thymus mastichina* (L) L. ssp. Mastichina, and *Thymus albicans*. Proceedings of the Congress on Medicinal Aromatic Plants. 1999;**501**:4

[22] Znini M, Cristofari G, Majidi L, Mazouz H, Tomi P, Paolini J, et al. Antifungal activity of essential oil from asteriscus graveolens against postharvest phytopathogenic fungi in apples. Natural Products. 2011;**6**(11):193-234

[23] Chafer M, Sanchez-Gonzalez L, Gonzalez-Martinez C, Chiralt A. Fungal decay and shelf life of oranges coated with chitosan and bergamot, thyme, and tea tree essential oils. Journal Food Science. 2012;**77**:182-187

[24] Chen CY, Zheng JP, Wan CP, Chen M, Chen JY. Effect of carboxymethyl cellulose coating enriched with clove oil on postharvest quality of Xinyu mandarin oranges. Fruits. 2016;**71**:319-327

[25] Ramezanian A, Azadi M, Ghalamfarsa R, Saharkhiz MJ. Effect of *Zataria multiflora* Boiss and *Thymus vulgaris L*. essential oils on black rot of 'Washington Navel' orange fruit. Postharvest Biology Technology. 2016;**112**:152-158

[26] Chen C, Cai N, Chen J, Wan C. Clove essential oil as an alternative approach to control postharvest blue mold caused by Penicillium italicum in citrus fruit. Biomolecules. 2019;**9**(5):197

[27] Davari M, Ezazi R. Chemical composition and antifungal activity of the essential oil of Zhumeria majdae, Heracleum persicum and Eucalyptus sp. against some important phytopathogenic fungi. Journal of Mycology Medicine. 2017;**27**:463-468

[28] Desam NR, Al-Rajab AJ, Sharma M, Mylabathula MM, Gowkanapalli RR, Albratty M. Chemical constituents, in vitro antibacterial and antifungal activity of Mentha essential oils. Journal of King Saud University. 2017;**54**:453-521

[29] Bhutia DD, Zhimo Y, Kole R, Saha J. Antifungal activity of plant extracts against Colletotrichum musae, the

postharvest anthracnose pathogen of banana cv. Martaman. Nutrition Food Science. 2016;**46**:2-15

[30] Boubaker H, Karim H, Hamdaoui A, Msanda F, Leach D, Bombarda I, et al. Chemical characterization and antifungal activities of four Thymus species essential oils against postharvest fungal pathogens of citrus. Indian Crops Production. 2016;**86**:95-101

[31] Jeyaseelan EC, Vinuja T, Pathmanathan K, Jeyadevan JP. Control of plant pathogenic fungi using organic solvent extracts of leaf, flower and fruit of Lawsonia inermis L. International Journal of Pharmaceutical and Biological Archives. 2012;**3**(4):783-788

[32] Sarkhosh A, Schaffer B, Varga SA, Palmateer AJ, Lopez P, Soleymani A. In vitro evaluation of eight plant essential oils for controlling Colletotrichum, Botryosphaeria, Fusarium and Phytophthora fruit rots of avocado, mango and papaya. Plant Protection Science. 2018;**54**:153-162

[33] Jhalegar JMD, Sharma JJ, Singh D. In vitro and in vivo activity of essential oils against major postharvest pathogens of Kinnow (Citrus nobilis × C. deliciosa) mandarin. Journal Food Science Technology. 2015;**52**(4):2229-2237

[34] Reimann S, Deising H. Risks of resistance development and search for new targets. Archives of Phytopathology Plant Protection. 2000;**33**:329-349

[35] El-Sayed HE, Ziedan ESH. Fumigation of peach fruits with essential oils to control postharvest decay. Research Journal of Agriculture and Biological Sciences. 2008;**4**(5):512-519

[36] Tzortzakis NG, Economakis CM. Innovative Food Science Emerging. 2007;**8**:253

[37] Batish DR, Singh HP, Kohli RK, Kaur S. *Eucalyptus* essential oil as a natural pesticide. Ecology Management. 2008;**256**:2166-2174

[38] Arrebola E, Sivakumar D, Bacigalupo R, Korsten L. Combined application of antagonist bacillus amyloliquefaciens and essential oils for the control of peach postharvest diseases. Crop Protection. 2009;**29**(4):369-377

[39] Kordali S, Kotan R, Mavi A, Cakir A, Ala A, Yildirim A. Determination of the chemical composition and antioxidant activity of the essential oil of Artemisia dracunculus and of the antifungal and antibacterial activities of Turkish Artemisia absinthium, A. dracunculus, Artemisia santonicum, and Artemisia spicigera essential oils. Journal of Agriculture Food Chemistry. 2005;**53**:9452-9458

[40] Robertson GL. Food Packaging: Principles and Practice. 3rd ed. Taylor & Francis; 2012

[41] Colín-Chávez C, Soto-Valdez H, Peralta E. Diffusion of carotenoids from monoand bilayer polyethylene active packaging into soybean oil. Food Packaging Shelf Life. 2014;**1**(2):170-178

[42] Colín-Chávez C, Soto-Valdez H, Peralta E, Lizardi-Mendoza J, Balandrán-Quintana R. Diffusion of natural astaxanthin from polyethylene active packaging films into a fatty food simulant. Food Research International. 2013;**54**(1):873-880

[43] Dietrich R, Ploss K, Heli M. Growth responses and fitness costs after induction of pathogen resistance depend on environmental conditions. Plant Cell Environment. 2010;**28**:211-222

[44] Manganyi MC, Reginer T, Olivier EJ. Antimicrobial activities of

selected essential oils against Fusarium oxysporum isolates and their biofilms. South African Journal of Botany. 2015;**99**:115-121

[45] Lambert RJW, Skandamis PN, Coote PJ, Nychas GJE. A study of the minimum inhibitory concentration and mode of action of oregano essential oil, thymol and carvacrol. Journal of Applied Microbiology. 2001;**91**:453-462

[46] Zhaoxin C, Dandan Z, Xuemei G, Yali L, Jingyi S. The role of essential oils in maintaining the postharvest quality and preservation of peach and other fruits. Journal Food Biochemistry. 2022;**46**(12):543-587

[47] Safari ZS, Ding P, Nakasha JJ, Yusoff. Controlling *Fusarium oxysporum* tomato fruit rot under tropical condition using both chitosan and vanillin. MDPI. Coatings. 2021;**11**(3):367

[48] López-Gómez A, Ros-Chumillas M, Buendía-Moreno L, Martínez-Hernández GB. Active cardboard packaging with encapsulated essential oils for enhancing the shelf life of fruit and vegetables. Frontiers Nutrition. 2020;7:275-287

[49] Santoro K, Maghenzani M, Chiabrando V, Bosio P, Gullino M, Spadaro D, et al. Thyme and Savory essential oil vapor treatments control Brown rot and improve the storage quality of peaches and nectarines could favor Gray Mold. Food. 2018;**7**:7

[50] Zanetti M, Carniel TK, Dalcanton F, Anjos RS, Riella HG, Araújo PHH, et al. Use of encapsulated natural compounds as antimicrobial additives in food packaging: A brief review. Trends Food Science Technology. 2018;**81**:51-60

[51] Buendía-Moreno L, Ros-Chumillas M, Navarro-Segura L, Sánchez-Martínez MJ, Soto-Jover S, Antolinos V, et al. Effects of an active cardboard box using encapsulated essential oils on the tomato shelf life. Food Bioprocess Technology. 2019;**12**:1548-1558

[52] Zhou D, Wang Z, Li M, Xing M, Xian T, Tu K. Carvacrol and eugenol effectively inhibit Rhizopus stolonifera and control postharvest soft rot decay in peaches. Journal of Applied Microbiology. 2018;**124**:166-178

[53] Ma BX, Ban XQ, He JS, Huang B, Zeng H, Tian J, et al. Antifungal activity of *Ziziphora clinopodioides* lam. Essential oil against *Sclerotinia sclerotiorum* on rapeseed plants. Crop Protection. 2016;**89**:289-295

[54] Rasooli I, Rezaei MB, Allameh A. Growth inhibition and morphological alterations of *aspergillus Niger* by essential oils. Food Control. 2006;**17**:359-364

[55] Sharma N, Tripathi A. Effects of Citrus sinensis (L.) Osbeck epicarp essential oil on growth and morphogenesis of aspergillus Niger (L.) Van Tieghem. Microbiological Research. 2008;**163**(3):337-344